Tropical Fruits : Diseases and Pests

Tropical Fruits : Diseases and Pests

Dr. ARUN ARYA
Reader in Botany
Faculty of Science
The M.S. University of Baroda
VADODARA–390 002
E Mail : aryaarunarya@rediffmail.com

KALYANI PUBLISHERS
LUDHIANA – NEW DELHI – NOIDA (U.P.) – HYDERABAD – CHENNAI
KOLKATA – CUTTACK – GUWAHATI – KOCHI – BENGALURU

KALYANI PUBLISHERS

Head Office

B-1/1292, Rajinder Nagar, **Ludhiana**-141 008 • Ph : 0161-2760031, 2745756
E-mail : kalyanibooks@yahoo.co.in **Website :** www.kalyanipublishers.co.in

Administration Office

4779/23, Ansari Road, Daryaganj, New Delhi-110 002
Ph : 011-23271469, 23274393 **E-mail :** kalyani_delhi@yahoo.co.in

Works

B-16, Sector-8, **NOIDA** (U.P.)

Branches

1, Mahalakshmi Street, T. Nagar, **Chennai**-600 017 • Ph : 044-24344684

Gopabandhu Lane, Behind Govt. Bus Stand, Badambadi,
Cuttack-753 012 (Odisha) • Ph : 0671-2311391

3-5-1108, Narayanaguda, **Hyderabad**-500 029 • Ph : 040-24750368

10/2B, Ramanath Mazumdar Street, **Kolkata**-700 009 • Ph : 033-22416024

Arunalaya, 1st Floor, Saraswati Road, Pan Bazar, **Guwahati**-781 001 • Ph : 0361-2731274

Koratti Parambil House, Convent Road, **Kochi**-682 035 • Ph : 0484-2367189

No. 24 & 25, 1st Floor, Hameed Shah Complex, Cubbonpet Main Road, **Bengaluru**-560 002

KPP N 22519 05

Reprinted, 2011
Second Revised Edition, 2017

Gautam Graphic Printers
(GEN-15-TFDP)

ISBN 978-93-272-7578-0

PRINTED IN INDIA

At Process World India, C-90, Sec. 63, NOIDA
and published by Mrs. Usha Raj Kumar for
Kalyani Publishers, New Delhi-110 002

Dedicated with regards to

Late Dr. S.N. BHARGAVA

(A Phytopathologist and Teacher par excellence)

May the rain-God send showers according to our desire,
May the medicinal herbs, bare plenty of flowers and fruits,
And enable us to prosper, to earn more, and to restore sufficiently.

निकामे निकामे नः पर्जन्यो वर्षतु फलवत्यो
न ओषधयः पच्यनतां योग क्षेमो नः कल्पताम् ।

यजुः २२ : २२

इच्छानुसार वर्षे, पर्जन्य ताप धोवें ।
फल–फूल से लदी हों, ओषध अमोध सारी ।
वे योग–क्षेमकारी, स्वाधीनता हमारी ।।

Let the earth and the water, the air and
The fruits of my country be sweet my GOD.

Let the homes and the marts, the orchards
And the fields of my country be fully my GOD
– Gurudeve Ravindra Nath Tagore

जीवैर्गस्तमिंद सर्वं आकाशं पृथिवी तथा ।
अविज्ञानाच्च हिंसंति तत्र किं प्रतिभाति ते ।।

The sky and the earth are full of living beings ;
Unknowingly they are inflicit with sufferings.
What do you think of it, Oh, Brahmin ?
– Mahabharata-Aranyaparva

FORWORD

I am very happy that Dr. Arun Arya has revised his book on **Tropical Fruits : Diseases and Pests.** Fruits are very important in human diet, as they are the major source of vitamins and micronutrients. India provides ideal climatic conditions for growing a wide variety of tropical as well as temperate fruits. The varieties of fruits, which are grown in India, are unmatched with any other part of the world.

Major constraints in improving the productivity of fruits in India, particularly of tropical fruits, are the pests and diseases. These constraints not only adversely affect the production but also impede export of fruits from India. There is a great potential of export of fruits from India, which can be achieved not only by improving the quality of fruits but also through judicious management of pests and diseases. It is in this context that the book written by Dr. Arun Arya is of topical interest. The book deals with various diseases caused by fungi, bacteria and viruses, and infestation of plants by insect pests. The management practices, included in the book are easy to follow and will certainly be useful in minimizing the losses caused by pests and diseases of tropical fruits in the country.

I am sure the book will be found very useful not only by the students and teachers of horticulture but also by the growers and various organizations dealing with the export of fruits from the country. I congratulate Dr. Arun Arya for his efforts in updating and including information about more fruits in the new edition of his book.

Anupam Varma

National Professor

IARI, New Delhi

PREFACE

Van Loon (1921) has rightly mentioned ''The history of man is the record of hungry creature in search of food''. All human efforts in that direction receive a great set-back because of the activities of fungi, bacteria, viruses, rickettsia, mycoplasma and spiroplasma etc. Moreover birds and insects play a havoc.

Plant pathogens are in direct competition with man when they affect the fruit crops. Advanced horticultural practices have produced conditions to have uninterrupted life cycles of phytopathogen throughout the year. Thus, the opportunities for a rapid build up of serious infestation is greatly enhanced. Therefore, plant protection becomes an equally essential input in increasing production of food, fruits and vegetables.

Students and research scholars have to work hard in various librariesand consult a host of books and journals to get disease symptoms and control measures which necessitate a book on diseases of fruits.

In order to keep this book to practical length I have selected ten tropical fruit crops viz., mango, grapes, guava, banana, ber, sapota, papaya, Indian gooseberry, bael and jamun. Chapters on nutritional value of fruits and bio-chemical changes during pathogenesis are added to give the importance to such studies.

In our country plant pathologists have concentrated on the study of diseases of fruits during field conditions. In the present book various post-harvest diseases are also included. Many birds and insect pests cause severe damage to fruit trees. Such pests are discussed in detail. Any how I do not claim that the book includes all the known diseases and pests.

I thank Prof. K.S. Bilgrami, Prof. H.C. Dube, Prof. R.S. Mehrotra, Prof. G.P. Agarwal, Prof. H.S. Sohi, Prof. Anupam Varma, Dr. P. Pitakpaivan, Bangkok, Drs. N. Shanmuganathan and R. Emmett, Australia, Dr. Mary Anna Sall, California, Prof. J.W. Eckert, California, Drs. Om Prakash and K.C. Srivastava, for their timely help in preparation of the text of this book.

I wish to express my deep sense of indebtedness and reverence to Late Dr. S.N. Bhargava and to Dr. Bihari Lal, Reader, Department of Botany, University of Allahabad who taught me the first lesson of Plant Pathology. Special acknowledgements are made to Maj. N.S. Parihar, former Head of the Department, Dr. D.D. Nautiyal, Head, A.U., Dr. G.L. Tewari, Dr. D.N. Shukla and Dr. Ravi Shanker Pandey, P.C.S. for their kind help. I thank Shri D.M.Mohite, for neat typing and Shri Ravindra Dindorkar for photographs.

I may be allowed to pay my tributes to the sacred memory of Baba Shankaranand, who nourished me with sweet mangoes and delicious guavas during my childhood and whose blessings had always been a source of inspiration for me.

At last I am thankful to Mr. Raj Kumar and various members on the staff of M/s Kalyani Publishers for taking keen interest in the publication of the book.

ARUN ARYA

EXPLANATION OF PLATES

Plate 1.

(*a*) Disease symptoms of Mango Malformation

(*b*) Die back disease caused by *Lasiodiplodia theobromae*

(*c*) Anthracnose of mango leaf

Plate 2.

(*a*) Aspergillus rot of Mango, symptom on stalk end

(*b*) A petridish showing culture of *Aspergillus niger* and *Fusarium sp.*

Plate 3.

(*a*) Aspergillus rot

(*b*) Colletotrichum rot

(*c*) Pestalotia rot

Plate 4.

(*a*) Colletotrichum rot of *Mangifera indica* var. Dashehari

Plate 5

(*a*) Colletotrichum rot on variety Totapari

(*b*) Infected fruits of mango due to *Phoma multirostrata*

Plate 6.

(*a*) Stalk rot of mango caused by *Lasiodiplodia theobromae*

(*b*) Rhizoctonia rot

Plate 7.

(*a*) Anthracnose of banana

(*b*) Pestalotiopsis leaf spot

(*c*) Cordana leaf spot

Plate 8.

(*a*) Panama disease or banana wilt

(*b*) Leaf spots caused by *Cercospora musae*, Sigatoka disease

Plate 9.

(*a*) Phomopsis rot of Ber

(*b*) Phoma rot of Ber

Plate 10.

(*a*) Anthracnose of papaya

(*b*) Acervuli of *Colletotrichum papayae* $\times$ 490

Plate 11.

(*a*) Fusarium rot of papaya

(*b*) Mycelium and conidia of *Fusarium oxysporum* × 490

Plate 12.

(*a*) A mixed infection of *Fusarium roseum* and *Colletotrichum* sp.

Plate 13.

(*a*) Fruit rot of papaya caused by *Thielaviopsis* state of *Ceratocystis paradoxa*

(*b*) Chlamydospores and endoconidia of *Thielaviopsis* × 950

Plate 14.

(*a*) *Aspergillus niger* rot of Aonla

(*b*) *Aspergillus terreus* rot of Aonla

Plate 15.

(*a*) Cladosporium rot of Aonla

(*b*) Phomopsis rot of Aonla

Plate 16.

(*a*) Aspergillus rot of Pomegranate

(*b*) Culture of *Aspergillus* sp.

Plate 17.

(*a*) Heart rot of pomegranate caused by *Aspergillus niger*

(*b*) Penicillium rot

Plate 18.

(*a*) Leaf infected with *Xanthomonas citri*

(*b*) Twig showing canker symptoms

(*c*) Aspergillus niger rot

(*d*) Sour rot caused by *Geotrichum candidum*

(*e*) Penicillium rot

(*f*) Aspergillus rot

Plate 19.

(*a*) Aspergillus rot of Litchi

(*b*) Conidial heads of *Aspergillus niger* × 940

CONTENTS

1

Mango

1.1. GENERAL ACCOUNT

It is called the king of fruits. The mango, a fruit that originated in India, has admirers all over the globe. It finds mention in the Vedas too. Truly, the mango is India's greatest contribution to the international sweet teeth. Mango undoubtedly deserves to be the national fruit of India. In area of production, nutritive value and popularity, no other fruit can compete with it. It occupies the same position in India as is occupied by the apple in temperate climates and grapes in subtropical areas. Its cultivation in India is estimated to more than 6,000 years old.

It has been held in high esteem by the Hindus. Its blossoms are used for the worship of goddess Saraswati and festoons (Toran) of mango leaves are strung over doorways on auspicious occasions. The mango has been described as 'Kalpavriksha' or wish-giving- tree. Its common vernacular name 'Aam' means 'the common'. It is not only common throughout India, but it is also the fruit of the common man.

Botanically the mango belongs to the family of Anacardiaceae to which also belong the cashewnut and pistachionut. The cultivated mango is scientifically called as *Mangifera indica* L. Wild mango forests of this species and *Mangifera sylvatica* are found in Assam. Another species *M. khasiana* is said to have existed in India, but has now become extinct. Other species of *Mangifera* such as *altissima, caesia, foetida, longipetrilata, maingayi, odovata, pentandria,* which are found in South-East Asia and the Philippines, are not found in India.

The names of fruit in other languages like mango in English, the mango or mangga in Malaya and Java and Mango-kwo in Chinese, is seem to have been derived from the Tamil name mankay or mangay. Its English name seems to have been derived through the Portuguese.

The mango had become established in India at a very early stage. The invading armies of Alexander the Great, found it established in the Indus Valley in 327 B.C. It is now a common fruit in all the Malay Archipelago, Indonesia, the Philippines, the West Indies and Madagascar. It has also been introduced into many other lands like Brazil, the U.S.A. and Queensland in Australia. It has become established in Somaliland on the estern coast of Africa before 1331. By the 16th century, it had reached the Persian Gulf. By 1690 it was grown under glass house conditions in England.

Mango is grown in atleast 87 countries (Prakash and Srivastava, 1987) but no where it is so greatly valued as in India where it occupies about one million hectare or 70 per cent of the area devoted to fruit crops with a total production of 8.21 million tonnes. India is the world's largest producer, far exceeding the leading countries. It contributes about 65 per cent of the total world production (Bonded, 1980). This fruit is reported by many authorities to be eaten by more people of the world than any other fruit. This does not seem reasonable but when one considers that India, the Indo-China area, Indonesia and the Philippines as well as small portion of China, have the highest population density in the world and the greatest mango production in the world from both cultivated and wild trees, the mango becomes an important fruit (Goldweber, 1968).

The mango is grown in almost all states of India. Besides well organised orchards it is grown as road side trees and as isolated tree in backyards of houses and on boundaries of the fields. Of the various states of India, U.P. has the largest area (33.2%) followed by Bihar (13.5%), Andhra Pradesh (13.4%) then comes Tamil Nadu, Kerala, West Bengal, Orissa, Gujarat, Maharashtra etc.

The climate of India is ideally suited for its cultivation and it is found growing in every nook and corner of the country except at altitudes above 3,000 feet. Usually the trees are neglected with no proper manuring, watering, and over crowding of trees. The tree assumes enormous proportions under ideal conditions. A classic example is the chhappar tree in village Buraid near Chandigarh in Punjab. This tree has a trunk with a girth of 32 feet and branches upto 80 feet long and 12 feet in circumference. It covers an area of 2,700 sq. yards and the average yield of the tree is 450 mds. The trees also live long, some trees are old as 300 years said to live even today.

A ripe mango is cut and eaten by people. On the other hand, juicy, sucking varieties are consumed in large quantities. The juice of mango fruit is dried in the sun in the form of solid sheets called mango leather or 'Aampapad' which is a sour-sweet product. The use of mango fruit starts from the very younger stage when the fruits are peeled and cooked with pulses. They are made into sour or sweet chutneys. Mango is also made into pickles, preserves, mango powder (Amchur) used in preparation of vegetables, and chutneys of all sorts. Mango pickle is the most important pickle and along with mango chutney constitutes the most important item of export of preserved fruit products from India. Mango-fool or milkshake, a refreshing drink is made by mixing mango pulp with milk and sugar. Except root every part of the plant is used. Even the Kernel inside the stone is not wasted. It is boiled and eaten or roasted and powdered and cooked into gruel. It contains over 8% protein. All mango kernals produced in India in a year are estimated to contain 70 million pounds of protein and 780 million pounds of starch. It is an outstanding source of vitamin 'A' and a good source of vitamin 'C', apart from the usual content of minerals and other vitamins.

Good mango varieties contain over 20 per cent of total sugars in which the non-reducing sugars are more than the reducing sugars. The acid content of ripe dessert fruit varies from 0.2 to 0.5 per cent and the protein content is about one per cent.

The mango is also considered to have some medicinal properties. The ripe fruit is fattening, diuretic and laxative. The smoke of burning leaves is supposed to cure hiccups and some throat troubles and the kernel is effective against diarrhoea and asthama. Baked and sugared pulp is given to patients of cholera and the plague. The bark is a source of resins and gum. The gum and resinous substance exuded by the stem end of the harvested fruit are mixed with lime juice and given in cases of scabies and skin infections. The bark is used in tanning of leather. The wood has many uses; it is used as timber for building houses. It is used in making packing cases and percussion drums. Its leaves are used as fodder. Thus the mango tree has several uses and are fully utilized in India.

Innumerable varieties of mango have been reported from India. They include types which are planted with seedlings progenies. In such plants the size, quality and colour of the fruit varies from one tree to another. They are not so well known as other commercial varieties, though about 2/3 of the total area is under such varieties which are commonly known as 'Deshi'. The number of commercial varieties is estimated at over 1,000, of which 30 are important.

Important varieties of U.P. and Punjab are Langra, Malda, Chausa, Fajri, Gulab and Krishan-Bhog, but Dasheri of Malihabad (Lucknow) is outstanding. It is a mid-season variety of delicate taste, flesh is fibreless, the fruit is elongated, dull yellow and medium sized. It has good keeping quality. The tree is small with regular bearing habit.

Bombai, Gulab khas, Himsagar, Krishan Bhog, Fazli, Sukul and Langra are the important varieties of Bihar. In Bengal, Bombai Bhutto is quite common.

In Western India the chief varieties are Alphonso or Hafus or Badami, Pairi, Kesar, Borsha, Fernadin, Rajapuri, while in Southern India, Bangalora or Totapuri or Thevadiyamuthi, Neelum, Rumani and Pairi are grown. In Andhra Pradesh, Bangaupalli or Safeda, Mulgoba and Suvarnarekha are important varieties.

Isozyme is an important tool to classify mango varieties. Work done at Bangalore shows that in 200 polymbryonic varieties, there are 34 alleles which show monomorphism and 19 show polyemorphism. It also shows that cultivars could be classified into 7 phenotypes. The North Indian mango varieties differ genetically from the South Indian varieties. The North Indian varieties among themselves have low genetic distance but had high genetic distance with the south Indian varieties.

According to IIHR's director, Dr. R.M. Pandey, the thrust is in three directions; First to develop red coloured mango varieties with an eye on export. The foreigners are apparently fond of the blush which looks attactive when the fruit is exhibited on super market shelves. Second, breeding is aimed at developing fruits resistant to spongy tissue, a physiological disease marked by the development of hard, dry white patches in the pulp. This cannot be detected from the outside. Third, a major aim of mango breeding at IIHR is to develop dwarf varieties.

According to Dr. C.P.A. Iyer (Indian Express, 7th May 1991) who is incharge of the Institute's mango breeding programme, the dwarf variety yield per hectare is higher than that of traditional varieties, because more mango trees can be accommodated in the same area. Also, as the mango requires more sunlight to develop properly, the small size of dwarf varieties, makes for increased exposure to the sun consequently, the trees bear more fruits which are also of better quality. Another equally important advantage of the dwarf variety is that it is easier to carry out plant protection operations on it. Obviously because it is easier to spray a small tree rather than a big one.

Effect of irrigation has profound effect in fruit drop. This phenomenon in 'Alphanso' variety was studied by Bhambid *et al.* (1988).

In Kualalumpur (Malaysia), Tarmizi *et al.* (1988) studied the effect of harvesting time on the incidence of fruit rot. They observed that 'Harumanies' mangoes harvested at different stages of maturity showed symptoms of insidious fruit rot whilst on trees in Karang early harvesting at 1-2 weeks before normal harvesting (90-100 d after fruit set) only slightly reduced the fruit rot incidence. Ethephon hastened ripening but did not improve fruit quality.

Compared to calcium carbide, Ethrel (2-chloroethyl phosphonic acid) dip treatment is safer and has been used to ripen physiologically premature mango fruits. Mango fruits of Dashehari harvested 2 weeks prior to harvesting maturity ripen to a better quality with dip treatment of 750 ppm Ethrel in hot water at 52 + 2°C for 5 minutes than calcium carbide treated with Ethrel produce uniform ripening and better colour development with lower concentrations of Ethrel (100-500 ppm) in hot water (Tandon and Kalra 2001).

Fifteen year old trees of the cultivar 'Dashehari' were subjected to 6 treatments involving leaf removal and ringing. Removing 50 or 100% of the leaves gave 79 or 73% fruit set respectively but when it was combined with shoot ringing levels of fruit set were 13 or 0% respectively. Ringing alone resulted in 19% fruit set, compared with 90% in the intact control (Khader *et al.*, 1989). Effectiveness of hot water treatment, waxing and cool chamber storage in prolonging shelf-life of mango was studied by Thangaraj and Irulappan (1988). They observed least weight loss 2-3% and longest shelf life 9.7 days in 'Neelam' variety, after hot water treatment at 52 ± 1°C + waxing with 10% frutox + cool chamber storage. In Australia, Brown (1986) dipped mature fruits in 0.5 g/1 Benomyl and 5% of Peerless citrus wax and stored at 20°C. The application of Peerless to springy (preripend) fruits extended the shelflife from 11 days (in untreated control) to 14 days with comparable quality.

Peterson (1986) reported various diseases of mango in Australia and their control measures as discussed in first Australian mango research workshop at Melbourne, Australia.

The fruit was treated with 6% wax emulsion (Waxol-w-emulsion) containing0.4% sodiumorthophenylphenate and dried in air current before packing (Ram and Date, 1971). The losses in weight of mangoes during the period under storage are significantly higher in the control. The skin coating thus reduced over all physiological losses and the treated fruit retained gloss for a period of 4 to 5 days and more than the corresponding control lot of fruits.

1.2. INSECT PESTS

1.2.1. The Mango Hoppers : *Idioscopus niveosparsus L., I. clypealis L. & Amritodus atkinsoni L.*

(Family—Cicadellidae : Order Hemiptera)

The species of hoppers attack mango inflorescence throughout India.

The damage is due to 1. heavy egg laying on the inflorescence stalk and florets which causes withering and drooping, and 2. adults as well as nymphs which pierce the tissues of flower panicle and suck the sap causing withering and shedding of flower. In case of severe attack honeydew excreted by the hoppers cover the inflorescence and leaves giving rise to development of sooty mould which hinders the setting of fruits.

Loss due to hopper varies from 25 to 60 per cent. The extent of damage caused may vary in different years.

I. niveosparsus is slightly smaller and has a prominent white bar crossing its dusky wings and three spots on the scutellum. It is the most injurious species. These are found all through the year hiding on the bark or leaves of mango till the trees put forth their blossoms. The insect appears in large numbers during cold weather (Nov. to Feb.) as soon as the flower buds begin to form. Eggs are inserted into the florets and stalk of the inflorescence and the eggs hatch in four to seven days. The nymphal period is 8 to 10 days. The total length of the life-cycle from egg to adult varies from 12 to 17 days. There may be two to three broods during the blossoming period of mango.

The Natural enemies of the mango hoppers include *Pyrilloxenos compactus* (Stripsiptera), *Epipyrops fuliginosa* (Epipyropidae) and the entomogenous fungi *Isaria stellata* and *Hirsotella versicolor*.

Control

Spraying DDT (0.1%) thrice at 10 day intervals may be necessary to control the pests.

Spraying carbaryl (0.1%), toxophene, endrin etc. have also been reported to be useful.

These suck the sap from the tender parts and secrete a sweet sticky substance on which sooty mould develops. It can be controlled by three sprays of 0.15 per cent Carbaryl or 0.4 per cent Monocrotophos (1 ml Nuvacron per litre of water) or 0.5% of Metacid (1 ml/litre of water) or Phosphamidon 0.05% (1/2 ml *Mango* 5 Dimecron/litre of water). The first spray should be given after the emergence of panicles but before flower opening, second when the fruit is pea-size and the third after 15 days of the second spray (Pandey, 1993).

1.2.2. The Oriental Fruitfly : *Dacus dorsalis* Hendel *or Chaetodacus*

(Family-Tephritidae : Order-Diptera)

This is one of the important pests of mango in India. Apart from mango it also attacks a wide variety of fruits and vegetables like *Solanum melongena, Psidium guajava, Citrus* sp.

etc. It attacks the semiripe fruits of mango and citrus and sometimes causes appreciable damage to them. The insect is light brown with transparent wings. The maggots feed on the pulp of the fruit and cause brown rotten patches on the attacked fruit which fall down eventually. The maggots come out of the fruit and pupate in the soil.

Control

In South India three rounds of malathion (0.1%) in combination with either DDT (0.1%) or Endrin (0.02%) or Fenthion (0.1%) at monthly intervals commencing $1\frac{1}{2}$ months after flowering, minimises damage by the pest.

1.2.3. Mango mealy bug : Drosicha mangiferae Gr.

(Family-Margarodidae : Order-Hemiptera)

It is a serious pest in Northern India, especially on mango, citrus and jujuba or ber.

The Nymphs and adults suck the sap from tender shoots, leaves and the fruit. In cases of severe infestation the tree shows sickly appearance, leaves and flowers are shed resulting in poor fruit setting and premature fruitfall. The honeydew excreted by the insect favours development of sooty mould which also affect normal photosynthetic activity. The older fruit do not develop properly.

The winged male, soon after emergence, mates with the apterous female. Though the female mates only once the male fertilises more than one female. The mated female descends the tree and reaches the soil for oviposition. Each female lays 56 to 580 eggs. The eggs are laid inside cottony white ovisacs from April to the beginning of June at a depth of 50 to 175 mm in loose soil near or away from the base of the trees. If the soil is hard the eggs may be laid in the cracks of the soil and bark.

The nymphs hatch out from the eggs during September to February, but mostly during November. The incubation varies from 98 to 185 days which is attributed to differences in soil temperature and humidity. The nymphs start ascending the trees from October to November. They collect in the axils of twigs and cracks of the bark in clusters of 17 to 283. The females have four nymphal instars and males three. Total developing period is about 250 days. The black ant *Camponotus compressus* attends on the bug for its honeydew excretion.

Natural enemies

Coceinella septumpuctata, Menochilus sexmaculatus, Rodoliafumida, larvae of *Chrysopa scelestis* and birds are predacecous on the insect. An entomogenous fungus *Aspergillus parasiticus* is parasite on it. A mite *Bochartia* sp. is an exoparasite on the adults and nymphs.

Control

Its eggs are found in the soil and should be destroyed by repeatedly raking up the soil from July to October. During monsoon the soil should be treated with 10% BHC (Ranjit Singh, 1969). The mealy bug is hard to kill using pesticides and the best control measure is to apply grease-coaltar (1:2) or *Namhar* band (Bindra, 1967) or it can be prepared by boiling together rosin (6 lb) and castor oil (5 lb) for 10 minutes. The castor oil can be replaced by rapeseed oil (2 lb). Other materials including some proprietary products can also be used (Ranjit Singh, 1969) in order to prevent the young mealy bugs form climbing the trees. The first band is applied in December which are effective for six to eight weeks. The band is renewed in mid January.

Application of aldrin (5%) dust around the base of trees and spraying of parathion (0.1%) or malathion (0.05%) for controlling early instar nyphs will be useful. Bindra (1967) recommended killing of insects near the band by heat from a blow lamp.

1.2.4. Mango mealy Bug : Rastrococcus iceryoides Gr.

(Family-Pseudococcidae : Order-Hemiptera)

It is a common pest of mango in Northern states. It is found throughout India. The nymphs and adult females suck the sap from leaves, tender terminal shoots, inflorescence and the fruit. The leaves are infested by the early instar nymphs. Infestation causes the shedding of flowers and the fruit setting is thus affected severely. Honeydew secreted by the mealy bug favours development of sooty mould.

The insect is abundant during April-June and thereafter declines. By October it is not generally observed. In Jabalpur it passes 6 to 9 generations in a year and over winters as adult female. It lays 400 to 630 eggs. The eggs are laid in a waxy ovisac on shoots and infloresceneces. The red tree ant *Occophylla snaragding* is found in association with this mealy bug.

Host Plants : *Mangifera indica, Boswellia* sp., *Hibiscus-rosa-sinensis, Gossypium sp., Lablab niger, Citrus sp., Ficus indicus* etc.

Natural enemies :

1. *Tetrastichus* sp. (Enlophidae) is parasitic.
2. The larvae of lycaenid butterfly *Spalgius epius* is predaceous on the mealy bug.

1.2.5. Fruit borers : *Hyalospila leuconeurella* Rag. and *Phycita umbratelis* H.

(Family-Phycitidae : Order-Lepidoptera)

H. leuconcurella and *P. umbratelis* are common fruit borer moths. *H. leuconeurella* bores the mango fruit in South India (Nayar *et al.*, 1981). Moths are small, delicate insects with elongate forewings having no R 5. The hind wings are broad and bear peeten of hairs on the dorsal side. Males have swollen basal antennal segment. Larvae feed on the fruit.

Control

The borer pest can be controlled by Endrin (2%), Carborl (0.1%) and DDT (0.25%), but spray of chemicals on fruit should be minimized.

1.2.6. The Mango Blossom Gall Midge : Procystiphora (= Dasineura mangiferae Mani

(Family-Cecidomyiidae : Order-Diptera)

Common throughout India on mango flowers. The larvae feed on the contents of floral buds and cause conical gall-like structures. They dry up in course of time and drop down.

The adult fly is orange coloured whihc lays 3 elongate cylindrical eggs between the calyx and corolla of the bud. The incubation period is 30 to 36 hours. The yellowish full grown larva pupates within the bud in a thin silken cocoon. Rarely 2 to 3 pupae may be there in a single bud. The empty pupa remains attached to the bud. Life cycle is completed in 12 to 14 days.

1.2.7. Red ant : Occophylla smaragdina Fb.

(Family-Fromicidae : Order-Hymenoptera)

The red ants which build their nests by stitching leaves together do not cause any damage directly, but besides being a nuisance, because of their painful bites they protect and spread certain scale insects for feeding upon their honey dew. The red tree ant is found associated with mango mealy bug (*Rastrococcus icervoides*).

Control

The insect can be controlled by removal and destruction of the nests and by spraying 0.15 per cent BHC of chlordane.

1.2.8. Mango Inflorescence Midge : Erosomyia indica Grover & Prasad.

(Family-Cecidomyiidae : Order-Diptera)

This is the most important of the midges that attack mango inflorescence in India. The insect is distributed throughout India. It attacks the inflorescence at the bud stage, elongating axis of the inflorescence and small developing fruits. The damage inflicted by the pest is often serious. It causes stunting of inflorescence and buds do not open. The infested fruit is pale and becomes shapeless and falls down when larvae jump out of them. In leaf mid-rib and petiole the galls are formed. The larvae feed on flower buds, inflorescence axis and newly set fruits. They become fully grown in 8 to 10 days. When there is shortage of food the larvae enter diapause in the soil at about 2 cm depth and remain in this stage till the next mango season, where they pupate and emerge as adults.

1.2.9. White ants : *Neotermes (Kalotermes) gardneri Synder and Trinervitermes biformis*

(Family-Termitidae : Order-Isoptera)

Termites, commonly but wrongly called white ants, may weaken and ultimately kill the fruit trees especially seedlings and young trees by gnawing at the roots. Consequently, gap filling may be necessitated. The attack is more serious in lighter soil and where under-compost organic matter is applied as manure. Besides mango, other fruit trees like jamun, grape vines etc., are also damaged.

Control

Application of 5 per cent Aldrin, BHC or Cholordane at the rate 1/4 kg per tree in the soil around the tree is recommended. It is best applied as a pre-disease treatment.

1.2.10. Mango blossom midge : *Dasineura amaramanjarae Grover*

(Family-Cecidomyiidae : Order-Diptera)

The insect occurs throughout India. It feeds on internal contents of buds which fail to open and drop down resulting poor fruit setting.

Mango flowering coincides with the emergence of adult midges. The preceding year larvae diapausing in the soil pupate with the mango flowering season and emerge as adults.

The midges mate soon after coming out and insert 3 to 6 eggs per bud. Every female lays 40 to 50 eggs. The incubation period is 30 to 36 hours. The orange red larvae come out of bud and fall down. The larval period is 10 to 13 days.

It pupates in the soil at 4 to 8 cm depth in a cocoon of silk and emerges as adults in 4 to 6 days. The larvae enter disabuse at the depth of 15 to 18 cm in the end of fruiting season.

1.2.11. *Acanthocoris scabrator Fabr*

(Family-Corcidae ; Order-Hemiptera)

Acanthocoris scabrator was reported as a new pest of mango fruit in India. It feeds on the tender mango fruits by piercing the outer rind and sucking the juice from inside. Exudation ooze out through the points and skin around these spots becomes dark.

1.2.12. Table : Some other Insect Pests of Mango

S.No.	Name	Family	Order
1.	*Amrasca splenden* Ghauri (Leaf hopper)	Cicadellidae	Hemiptera
2.	*Aleurocanthus mangiferae* Q. & B. (Mango aleyrodid)	Aleyrodidae	Hemiptera
3.	*Laccifer lacca* Kerr.	Lacciferidae	Hemiptera
4.	*Pulvinaria cellulosa* Gr.	Coccidae	Hemiptera
5.	*P. psidii* Mask	Coccidae	Hemiptera
6.	*P. polygonata* Ckll.	Coccidae	Hemiptera
7.	*Diaspsis barberi* Gr.	Diaspididae	Hemiptera
8.	*D. cinnamomii-mangiferae* Newst.	Diaspididae	Hemiptera
9.	*D. mangiferae* Newst.	Diaspididae	Hemiptera
10.	*D. rosea* Bouche	Diaspididae	Hemiptera
11.	*Lipidosaphes gloverii* Rack	Diaspididae	Hemiptera
12.	*Rhipiphorothrips cruentatus* Hood (Leaf thrip)	Thripidae	Thysanoptera
13.	*Scirtothrips dorsalis* Hood (Inflorescence thrip)	Thripidae	Thysanoptera
14.	*Anaterachyntis simplex* Wlsm. (Inflorescence caterpillar)	Cosmopterygidae	Lepidoptera
15.	*Indarbela tetraonis* M. (Bark borer)	Gracillariidae	Lepidoptera
16.	*Argyroploce aprobola* Meyr. (Leaf roller)	Eucosmidae	Lepidoptera
17.	*Alassomya tenuispatha* Kief (Leafgall)	Cecidomyiidae	Diptera
18.	*Amradiplosis amraemyia* Rao (Gallmidge)	Cecidomyiidae	Diptera
19.	*Batocera rufomaculata* Deg. (Mango stem borer)	Cerambycidae	Coleoptera

DISEASES CAUSED BY FUNGI

1.3. MANGO MALFORMATION

The malformation disease is a serious threat to mango industry in several mango-growing areas of the world. It does not result in any malformation of the mango fruit as may be

Fig. 1

Fig. 2

Fig. 1, 2, 3 Mango Malformation

connotated from its name, but produces abnormal vegetative shoots and inflorescences which do not bear any fruit. And hence it is also called unchy top. The disease is caused by a fungus *Fusarium moniliformae* (Summanwar *et al.*, 1966, Varma *et al.*, 1972, 1974, Varma, 1983). It was first observed in the Indian subcontinent (Watt, 1891 ; Burns, 1910). Since then it was also been found to occur in the Middle East (Attiah 1955, Hassan, 1944 ; Hifny *et al.*, 1978), South Africa (Schwartz, 1968), Israel, Central America, Mexico, and the U.S.A. (Malo and McMillan, 1972).

1.3.1. Symptoms

Mango malformation (MM) induces abnormal development of vegetative shoots and inflorescences the two phases of the disease commonly referred to as vegetative and floral malformation.

The vegetative malformation is more pronounced in young seedlings and seedling trees than in the grafted plants. The grafted seedlings develop excessive vegetative branches which are of limited growth with very short internodes. These dwarf branches form bunches of various sizes which are often produced on the top of young seedlings giving a bunchy top appearance.

The axillary buds of dwarf and even normal looking branches are usually enlarged indicating disturbance in apical dominance. These buds persist even when dry and due to their crowding at nodes, form girdles around branches. Such swollen buds, referred to as scars, provide good indication of MM although not necessarily present on all affected branches.

The term floral malformation gives an impression that the flowers on affected panicles are abnormal and develop monstrosities, but actually there is little difference between normal and abnormal flowers (Varma *et. al.*, 1969). The term is used to distinguish normal and affected inflorescences which, indeed assume abnormal shapes due to thick, fleshy and profusely branched panicles. At the early stages of development, diseased and healthy inflorescences are difficult to differentiate. Flowers borne on malformed panicles are slightly larger than normal flowers and bear scanty pollen. Percentage of perfect flowers is smaller on malformed panicles than on normal panicles (Majumdar and Sinha, 1972). The affected inflorescences are of three types—heavy, medium and light (Varma *et. al.*, 1969). The panicles of heavy type are very compact due to excessive crowding of flowers and, unlike normal panicles, keep growing to form large hanging masses of flowers. Most of these dry up subsequently and hang as discoloured branches but some continue to grow till the next season. Medium types of malformed panicles are slightly less compact, and like the heavy types these persist on plants longer than the normal panicles (Singh and Chakravarti, 1935). The light types are only slightly more compact than the normal panicles and do not persist on plants. The malformed panicles give leafy appearance due to larger bracts than in normal panicles and not due to phyllody.

Malformed panicles, normally do not bear any fruit as the fruits set in the main season do not grow more than pea size. However, these have been observed to yield normal fruits during off season. The light types of var. Bombay Green also retains the fruit even in normal season.

The root system is also affected. Malformed seedlings usually have shallower root system with fewer tertiary roots and often the tap roots are twisted and necrotic (Varma *et. al.*, 1969).

The extent of infection varies from tree to tree and branch to branch in the same tree suggestive of localized infection. But, Varma *at al.* (1969) observed scars even on old branches of severely affected plants and when made to sprout under aseptic conditions the new shoots were made typically malformed indicating systemic infection in branches. All the branches produced on a malformed branch may not bear malformed panicles and even in a malformed panicle some secondary rachis may be normal (Malik, 1963). Conversely, on a healthy panicle some parts may be malformed.

1.3.2. Anatomical changes

Very little information in available about the anatomical changes due to malformation. Narasimaham (1954) found considerable transformation in the cortex and stele of panicles by the development of hyperplastic cells. Malik (1963) noticed pith cells of malformed panicles to be deformed and packed with starch. Pith of vegetative shoots had large number of spaces lined up with elliptical cells and contained large amounts of brownish fluid, which was unaffected by dilute chromic hydrochloride acid. A similar fluid was also found in the hypodermal cells of shoots and pedicels. The latter had dense epidermal hair and the epidermis was warty and puckered. No difference was found in the conducting tissue of shoots and panicles. Conducting tissue of the leaves was poorly developed on affected branches.

Histopathological and histochemical studies on mango shoots infected with mango malformation agent *Fusarium moniliforme* were conducted by Babu and Rao (1998). They found that during winter months the apical and axillary buds developed into reproductive buds giving rise to panicles. In the remaining seasons the buds invariably developed into vegetative shoots. The pathogenic interaction with some of the floral buds resulted into high incidences of malformation disease, which started early in floral season and extended upto April and reestablished in November. The fungus *Fusarium moniliforme* was often noticed associated with the floral and vegetative apices. The site of infection was marked by infusion of phenolic contents and proteins in the cells. Deposition of starch in pith cells of shoot was relatively more in the floral season. Infection by parasite alters the metabolism of tissues, so that respiratory substrates move towards the site of infection from elsewhere in the plant. Alternatively, more of these are produced by cells at the site of infection.

1.3.3. Etiology

Etiology of MM disease has been controversial. In the absence of sufficient information MM was earlier considered to be physiological, viral or acarological by different groups of workers. Relatively recent findings have, however, clearly demonstrated that the disease is caused by a fungus. As fungal etiology of the disease has not been confirmed from all the places where MM occurs, possibility of more than one incitant cannot be ruled out.

PHYSIOLOGICAL : Maries, who was the first to observe MM (Watt, 1981), felt that excessive soil moisture was the cause. No cure has been obtained when the plants were treated with different metallic and non-metallic compounds, however, application of iron chelates, seem to reduce the incidence malformation (Abo-El- Dahab, 1975 ; Minessy *et al.*, 1975) but the actual mode of action of iron chelates on MM is not known. The diseased tissue contains more carbohydrates and nitrogen than normal tissue (Malik, 1963 ; Tripathi, 1955). High C/N ratio may be responsible for greater percentage of staminate flowers in malformed panicle (Khan and Khan, 1962). Change in C/N ratio by applying increasing levels of nitrogen was found to decrease incidence of malformation by some (Khan and Khan, 1958 ; Prasad *et al.*, 1965) but not by others (Bindra and Bakhetia, 1971).

A critical look at the disease clearly indicates suppression of apical dominance, which is similar to that induced by morphactin (Chacko *et al.* 1973). This is further shown by reduction in floral MM by single application of 100-200 ppm NAA (α-naphthyl acetic acid) (Majumdar *et al.*, 1970 ; Shant, 1975).

VIRAL : Virus-like symptoms of the disease and failure to isolate any pathogenic organism lead to speculation (Sattar, 1946 ; Singh and Jawanda, 1961) about the viral nature of disease. There are some reports of successful transmission of the disease by grafting/budding (Bindra and Bakhetia, 1971).

ACAROLOGICAL : A phytophagous mite, *Eriophyes mangiferae*, also known as mango bud mite, is commonly found associated with mango plants including those affected by MM. These mites were shown to induce MM in test plants when transferred from diseased (Singh *et al.*, 1961) or even healthy source plants (Nariani and Seth, 1962). Successful reproduction of the disease does indicate that mites are involved, but it is uncertain whether these are vectors of some pathogen or of direct cause itself, as when mites are collected even from apparently healthy plants or branches growing in the open, they may carry a pathogen. Indeed, later findings have demonstrated the role of mites as vectors. Narasimhan (1954) considered mango mites to be the direct cause of MM as he found them in great umbers in the meristematic and tender regions.

FUNGAL : *Fusarium moniliforme* Sheld has been consistently isolated from various parts of MM affected plants in Brazil, India and U.S.A. The fungus, identified as *F. moniliforme* var. *subglutinans* (Varma *et al.*, 1974b), is characterized by shades of white aerial mycelium appearing powdery due to micro-conidia. Both vegetative and floral malformation can be induced in test plants by artificial inoculations of aerial branches with fungus. The fungus is mostly intercellular agglomerates in the cortex and phloem regions.

1.3.4. Control

The disease control has two distinct aspects (*a*) care of existing malformed plants (*b*) prevention from fresh infection. Tripathi (1955) and Singh *et al.* (1961) tried to control the disease by application of micro and macro nutrients without success. The best reduction of flower malformation was achieved by the higher N rate with Zn and Mn, each at 0.3 per cent in May and again in September (Azzouz and Dahshan, 1985). Application of 100 or 200 ppm of NAA has been reported by Majumdar *et al.* (1970). A foliar spray of urea at monthly interval during winter was considered useful in reducing the incidence of the disease (Khan and Hamid, 1958). Abo-El-Dahab (1975) reported that soil application of chelated iron gave good curative results.

Prunning has been found to be effective for curing malformation. According to Desai *et al.* (1962) spraying of pruned trees with insecticides was of no use over prunning alone but spraying or swabbing with Diaginon or Phosphamidon alone has been reported to be effective in reducing the malformation (Singh 1956, 1957) and in obtaining the normal growth from diseased terminal buds (Rai and Singh, 1967).

Varma *et al.* (1971) tested 34 fungicides, 17 insecticides and four growth regulators and found Benlate, Brestan, Busan, Captan, Dithane M-45, Panogen and Thiram were fungicidal at 1000 ppm. Aphidan and Phosphamidon were the most effective among the insecticides. Varma *et al.*, (1971) realized the need of systemic fungicide for the curative control of disease as the causative fungus is located in the cortex-phloem portion.

It is evident from the cited literature that at present no definite control measure for MM can be advocated. However, the following measures may be tried to reduce the incidence.

1. It is advisable to avoid bud wood from trees bearing malformed inflorescence for budding and grafting ;

2. Any observation of suspicious disease material should be reported to the nearest Horticulture centre.
3. Only certified plants should be used for propagation and further distribution to the various agencies within or outside the country.
4. As soon as the disease appearance is noticed, such material should be pruned and burnt.
5. Broad spectrum systemic pesticides should be used.
6. In area where more than 25 per cent plants are affected, deblossoming at bud burst stage should be advocated just to delay the flowering, (Varma, 1983).
7. Healthy orchards located in disease prone pockets should be spreadwith pesticides as a prophylatic measure to avoid further recurrence of disease.
8. Spraying of 200 ppm NAA in the first week of October is advocated followed by de-blossoming at the bud burst stage.

1.4. POWDERY MILDEW

The disease was first recorded on mango in 1914 in Brazil and the fungus was named and described by Berthet (Briton Jones, 1923 ; Uppal *et al.*, 1941). Powdery mildew caused by *Oidium mangiferae* Berthet, is one of the most important fungal disease and has been reported from almost all the mango growing areas of India, *viz.* Uttar Pradesh, Mumbai, Maharashtra, Andhra Pradesh, Karnataka and Tamil adu. The losses caused vary from 5 to 20 per cent and largely depend upon favourable weather conditions (Cheema *et al.* 1954). The mildew is a serious malady in the hill valleys and 'Terai' regions of Uttar Pradesh (Bose, 1953) and in Bombay state (Uppal, 1937).

Fig. 4. Powdery mildew on leaves.

The disease is quite widespread in Mysore state on variety 'Pairi', which is extensively cultivated throughout South India (Sridhar and Sohi, 1973). The infection first appears on the leaves in the month of January. Subsequently it attacks the tender twigs, inflorescence and the young fruit. The attack is usually very severe during the months of February to April resulting into defoliation and fruit shedding thereby affecting the yield adversely. New leaves are attacked mostly on their underside, but in advance cases both sides of the leaf are attacked. The symptoms are restricted to the area of the central rib. Such leaves curl and are distorted (Palti *et al.*, 1974).

Control

The most effective method of control is by dusting the plants with fine sulphur (250-300 mesh) at the rate of 1-3 lb of sulphur/tree (Cheema *et al.* 1954). The first application may be made soon after flowering and second 15 days later followed by a third one. Spraying with Guesarol (405.50) has been found successful in Tamil Nadu, Andhra Pradesh and Maharashtra for the control of the disease along with hoppers (Sohi, 1983). Out of a number of Wettable sulphurs and systemic fungicides tested at Indian Institute of Horticultural Research, Bangalore, maximum control was achieved with Cosan (0.2%) or Benlate (0.2%) when sprayed at fortnightlyintervals during January to May (Sridhar & Sohi, 1973). Karathane (0.05%) sprays were found effective in Uttar Pradesh and Maharashtra. Calixin and vigil were found effective in Haryana (Gupta and Dang, 1980). The best control was given by Topsin M-70 followed by Karathane and Bavistin among the six fungicides tested by Datar (1986). 3 sprays at 15 days intervals of either Microsul (0.2% wettable S) or Bavistin (0.1% carbendazim) controlled *O. mangiferae* satisfactorily (Prakash and Raoof 1985).

In trials conducted in South Africa, mono potassium phosphate (KH_2PO_4) was applied in a spray to flowering Kent and Tommy Atkins mango trees, either alone or with conventionally used curative fungicides *viz.* Punch (Carbendazim + flusilazole), Benlate (Benomyl), Folicur (tebuconazole). Unsprayed trees served as control. In Kent, the fungicide/mono potassium phosphate mixes were almost as effective as Bayfidan at full-strength against *Oidium mangiferae* (Oostheryse 2000). Field trials with 0.5 & 0.1% Triadimefin in Karnataka showed promising results (Desai 1998a). Ihsan *et al.* (1999) studied incidence of powdery mildew in Pakistan and found losses upto 70%. The disease was controlled by triadimenol (Bayfidan), tridemorph (Calyxin), fenarimol (Rubigan) and triadimefon (Bayleton).

1.5. LEAF BLIGHT

Averna Lacca (1922) from Brazil reported a species of *Macrophoma* causing black rot of mango fruit and trees. Patel *et al.* (1949) described a species of *Macrophoma* on branches and the fruit of mango which has been earlier considered a species of *Phoma* by Kanitkar and Uppal (1939). The disease was reported by Hingorani and Sharma (1956). It has subsequently been reported from Punjab and Maharashtra (Desai and Patel, 1963). The disease is caused by *Macrophoma mangifera* Hingorani and Sharma.

Spots gradually enlarge discolouring the surrounded tissue, which first become biscuit coloured, then brown and dark brown with slightly raised and broad dark purplish margin. Later they become ashy coloured, due to appearance of pycnidia. Spots are rounded in the beginning, later become oval or irregular depending upon the environmental conditions such as humidity and temperature etc.

The spores that are washed down by rains accumulate as droplets at the tip of the leaf, which dries up entirely and the infection travels downward toward the petiole sometimes half or more than half of the leaf may get involved. Pycnidia make their appearance mostly on the under-surface of the leaves, though sometimes on the upper surface too, as black dots scattered all over the necrotic portion. Dots are minute mostly single and innumerable. Infection is observed mainly on leaves and rarely on stems.

On stems, the lesions are elliptic which later girdle the stem at the point of infection. On fruit, water soaked, circular lesions are produced which enlarge rapidly and cause rot, particularly in storage. In field, however, infection is rare. Desai & Patel (1963) found that out of 38 varieties, Khandeshi borasio and Asal dadamio were resistant.

Control

The disease can be checked by spray of Burgundy mixture, Perenox and Dithane. Hingorani *et al.* (1960) found that spores of this pathogen failed to germinate in 2 per cent Burgundy mixture, 1/2 per cent Perenox and 1:9000 dilution of Dithane.

1.6. PHOMA BLIGHT

A widespread disease of mango discovered on cultivar Deshehari in the vicinity of Malihabad, Kakori and other adjoining areas (Prakash and Singh, 1977). It is now prevalent in West Bengal, Bihar, Maharashtra, Karnataka, Rajasthan, Gujarat, U.P., Punjab, Haryana and Goa (Anon, 1974-85). The disease is caused by *Phoma glomerata* (Corda) Woll and Hochapf. Symptoms of this disease are noticeable on old leaves only. Initially the lesions are minute irregular, yellow to light brown scattered over the leaf lamina. As the lesions enlarge, their colour changes from brown to cinnamon, and these become irregular. Fully developed spots are characterized by dark margin and dull grey necrotic centres. When infection is severe, the spots coalesce to form 3.5-13 cm patches, which result in withering and defoliation of infected twigs. Such infected trees can be located from a Benomyl (0.2%) and Miltox (0.3%) have been found effective (Prakash 1970, 1979).

Fig. 5. Infection on leaves due to phoma glomerata.

1.7. SCLEROTIUM ROT

The disease is known as Sclerotium rot of mango seeds and seedlings. In Philippines about 10 per cent mango seedlings died of stem rot and many of the seeds (Pyrene) rotted before or in the course of germination due to *Sclerotium rolfsii* and *S. delphinii* (Palo, 1933). Prakash and Singh (1976) found basal rot and rotting of stones in the nursery of Central Mango Research Station, Lucknow due to *S. rolfsii*. Gadre (1979) reported it from Maharashtra.

The disease is characterised by the presence of mycelial weft on the base of the stem at the ground level in patches. Beneath the mycelial growth, a dark brown spot may develop which gradually encircles the base of the stem. At this stage the seedling drops down due to loss of turgidity in the cells and seedlings dies. A vigrous growth of mycelium and selerotial bodies are seen in the stem and root zone of the infected tissues. Such roots are soft and black and the sclerotia of the fungus are attached to the cotyledons. The disease also causes severe rotting of the seeds during or before the germination.

Control

1. Cultural practices consists of use of less water during germination innursery beds.
2. Use of sanitary measures and healthy seeds.
3. Drenching the soil with corrosive sublimate (1 : 2000) (White, 1931).
4. Two minutes dipping of stones in Agallol/Brassical/Captan/Thiram and subsequent soil drenching at 10-15 days intervals to reduce the intensity of the disease.

1.8. DRY ROT OF SEEDLINGS

The dry rot of mango seedlings is caused by *Fusarium solani* (Mart) Sacc., (Kore and Patil, 1985). It was observed in Horticulture Farm, College of Agriculture, Parbhani (Maharashtra) in October-November 1979. The disease was 60 per cent in stone grafted and 20 per cent in seedlings.

Colour of young leaves became light green and finally light yellow and successively progress towards other leaves from tip downwards.

Purukayastha (1968) found *F. solani* on branches and trunks of mango trees which supports its prevalence in different parts of dry rotted grafts of mango seedlings.

1.9. ROOT ROT AND DAMPING OFF

Root rot and damping off is a serious malady of mango. It is caused by *Rhizoctonia solani* Kuhn. Gupta and Srivastava (1975) reported leaf spot disease of mango caused by *R. bataticola* (*Macrophomina phaseolina*) from India. This disease has been noticed in the nursery.

After the emergence from soil the leaves drop down. The stem is weak at this stage. During prolonged rainy and humid weather, infection occurs generally at or below the ground level with circular to irregular water-soaked patches. These patches enlarge and encircle the whole stem. On account of rotting, the diseased tissues become soft, dark brown or black and the entire seedlings collapses and dies. Rotting may spread above and below the stem down to roots. The roots get disintegrated.

Control

Soil treatment with Cersan is recommended. Bordeaux mixture (1.5%) should be sprayed on plant and soil.

1.10. WILT

Young mango seedlings died in random fashion in the orchard in Florida (Mariatt *et al.*, 1970).

The first external symptom of the disease is yellowing of leaves which have wilted slowly in scattered areas. One or more main branches die and the leaves turn brown, while remaining attached portion, caused a burnt appearance. The wood of affected branches shows brown discolouration and severely diseased plants die.

Control

Injection of Sequestene 138 (2 oz/5 gal water per tree) into the soil (Goldweber, 1976).

1.11. DIE BACK

The term 'die back' refers to death of plant from top downward. The disease is known from various mango growing areas of world (Prakash and Srivastava, 1987). In India it has been first described by Das Gupta and Zachariah (1939). Full account of this disease has been published by Prakash and Raoof (1985).

The disease is caused by *Botryodiplodia theobromae* Pat. The fungus is aggressive, vigrous and universal in distribution.

The disease is most serious in October-November just after the rains. It is characterised by dying back of twigs from top downwards particularly of the older trees followed by complete defoliation which gives an appearance as it has been scorched by fire. Discolouration and darkening of the bark at a certain distance from the top is external evidence of the disease. Such dark patches are generally seen on young green twigs and are rarely visible on older branches. Dying of young twigs begins from the base. Affecting the leaf midribs and extending outwards along the veins of the leaf edges. The upper leaves lose their green colour and gradually turn brown. Internal browning in the wood tissue is observed on splitting along the long axis. Cracks appear on woody stem from which gummy exudates come ot before they die out. When the graft union of young plant is affected it usually dies. It has been found that the infection occurs at node at variable distance below growing point and the part of the twig along and below the point of infection dies. Disease is also noticed in flowers and mangoes. The disease is caused by *Botryodiplodia theobromae* Pat, which is also mentioned as *Diplodia natalensis* Ev. by Su (1934). *Diplodia racifensis* causes 'Recife sickness' similar to die back disease in Brazil (Batista 1947). Bose *et al.* (1973) reported the disease to be caused by *Gloeosporium mangiferae.*

Histopathologoical observations reveal the diffused internal discolouration which appears as a dark streak between the stele and the cortex. The internal discolouration was manifested by browning of certain region of the cambium and phloem, where some of the cells were plugged with yellow gum like substance. Some hyphae of fungus may be present in xylem tissue. In advanced stages the cells of different tissues of stem were badly shrivelled.

Control

1. The disease can be controlled by clean cultural practices *viz.*, selection of healthy scion, sterlization of the budding knife, keeping the grafted tree in a relatively dry atmosphere and gradual exposure to full sun light.
2. Spraying the trees with copper oxychloride sulphate and applying the copper oxychloride paste at cut ends of them prevents the spread of disease.
3. Batista (1947) recommended pasting of trees with a mixture of oil and 5 per cent phenol.
4. All the openings of the insects tunnel should be sealed with paraffin wax to prevent the movement of beetle.
5. Srivastava and Tandon (1961) found Captan effective against the
6. pathogen *in vitro*.
7. Bordeaux mixture (5 : 5 : 50) or spray of copper oxychloride is most efficacious in controlling the disease (Prakash and Raoof, 1985).

1.12. STEM CANKER OF MANGO

Arya *et al.* (1987) reported occurrence of this disease in various orchards of Allahabad, Banda, Fatehpur, Kanpur and Pratap Garh in (U.P.) The disease is caused by *Botryosphaeria dothidea* (Maug. ex Fr.) Ces and de Not = *B. ribis* Grossenb and Dugg.

The first evidence of infection was shown by the presence of occasional blighted twigs and branches, leaves wilted, dried without defoliation. Elongated depressed lesions girdled the branches below the dying parts. Occasionally gummy substance attractive to ants accumulated at the surface. On the twigs and larger branches numerous cankers developed. The bark in the cankered areas became dry and cracked, and sometimes the wood was exposed in sharply demarcated cankers. Smaller cankers were irregular in outline, cratiform and approximately 1-2 cm in length. Dark brown to black coloured spots developed at the point of artificial inoculation which increased in size, covering the entire area of 'juncture'. Vascular elements of these infected twigs showed brown discolouration. At later stages cankers were observed on twigs having dried leaves.

In most of the orchards surveyed 10-15% trees were affected showing cankered stems and blighted twigs. None of these plants was, however, killed.

1.13. ANTHRACNOSE

The disease includes blossom blight, peduncle blight, leaf spot, twig blight, wither tip, fruit russeting or staining and fruit rot (Singh 1960). The disease is most serious on the fruit. It has been reported from several countries. In India this disease was reported by Mc Rae (1924) and Stevens and Pierce (1933). Uppal *et al.* (1935) observed that in Bombay it was due to *Gloeosporium raciborskii.* P. Henn, later Sattar and Malik (1939) recorded it from Punjab and identified the causal organism as *Glomerella cingulata* (Ston.) Spauld and Shrenk (*Colletotrichum gloeosporioides* Penz.)

Numerous oval or irregular vinaccous brown or dark brownish spots of variable sizes roundish or angular, scattered over the leaf surface are the characteristic symptoms. The spot may begin at the apex of the leaf or any other portion of leaf. Under damp conditions the spot grows rapidly forming elongated brown spots which when old become ruptured and blighted. They do not become much larger as the leaf grows, but often become dry and fall out giving the 'shot hole' appearance. Young leaves are more susceptible than older ones (Stevens, 1936). Vidhyasekaran and Durairaj (1973) reported role of auxins and phenol complex in shot hole syndrome development.

The petiole when affected, turn grey or black and the leaves droop down, become dry and ultimately fall off, leaving a black scar on the twig (Sattar and Malik, 1939).

The disease produces elongated black necrotic areas on the twigs. The tip of very young branches are attacked first and then twigs go on drying from the tip downwards. Under humid conditions all the branches of young plants dry. Young branches are affected on older plants, but the big branches are not attacked (Sattar and Malik, 1939). When the graft union is affected the plant usually dies. Gummosis is usually the after effect of the disease (Bose *et al.*, 1973).

Blossom blight and peduncle blight prevents the fruit from setting. On the inflorescence the characteristic symptoms are blackish brown speeks on the peduncle and flowers. Small black spots appear in the open flower panicle, which gradually enlarges and often coalesee to cause death of flowers. The whole flower stalk may become blackened, blighted and set no fruit. The infected flowers fall off. The severity of the disease may vary according to the prevailing weather condition (Prakash and Raoof, 1985).

The fruit is attacked during any stage of its development. The young fruit becomes infected and falls off in large numbers. The older fruit develops irregular small spots which spoil the whole fruit. Later on these spots enlarge and cover the entire fruit. Deep cracks develop deep into fruit and keeping quality of fruit is lost. Under moist conditions, the blackened areas become covered with minute pinkish pustules. These are the reproductive bodies of the fungus that produce large number of minute spores each of which is capable of causing fresh infection (Anon, 1974-85). On the ripening fruit the symptoms include sunken, blackish brown blotches upon which salmon buff masses of spores develop. In storage it causes severe rotting of the fruit. Rath and Mohanan (1986) observed black irregular lessions at stalk end.

The latent infection is carried from the field and develops further. Healthy fruits develop infection after coming in contact with the diseased fruit (Baker *et al.*, 1940 ; Sohi *et al.*, 1973). The later infection of the fungus does not begin to spread until ripening.

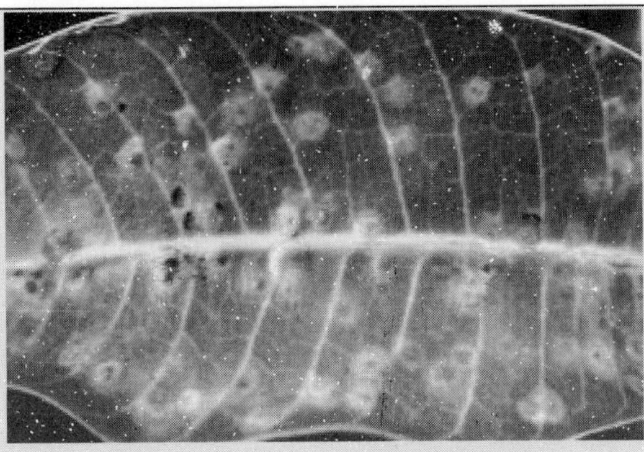

Fig. 6. Mango Anthranose leaf spot.

Fig. 7. Mango Anthranose leaf spot.

Fig. 8. Alternaria leaf spot on mango.

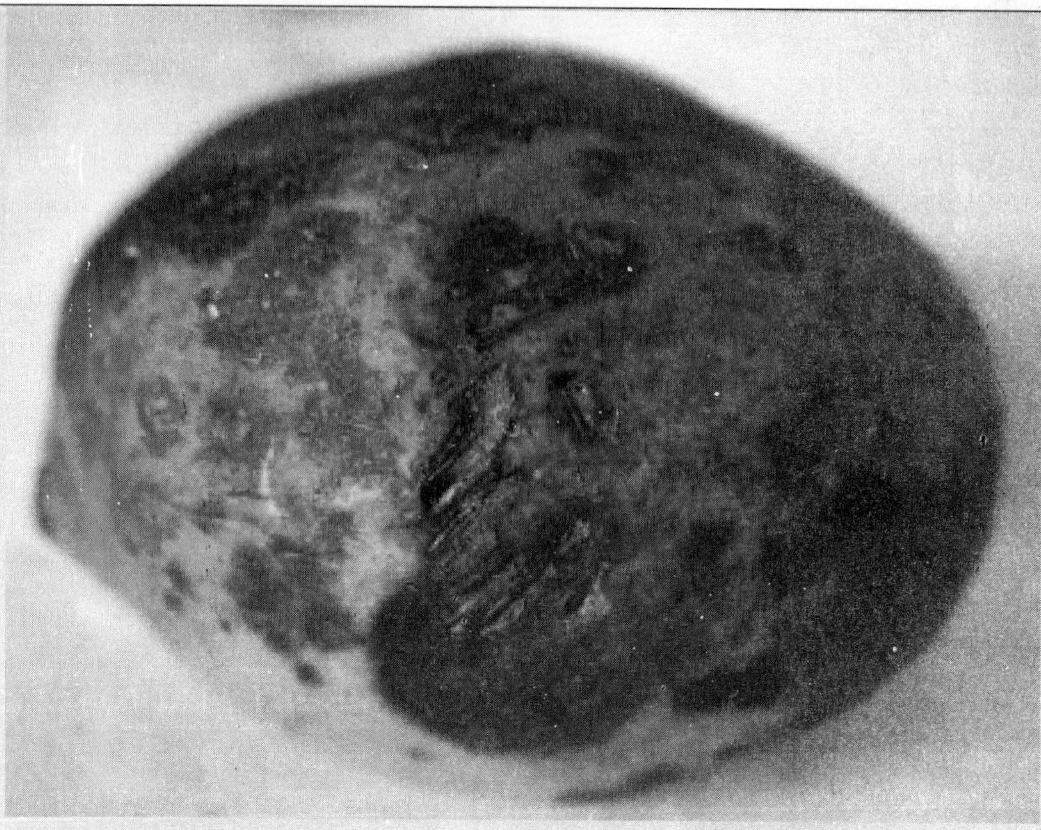

Fig. 9, 10, 11. Colletotrichum rot.

The causal organism is *Colletorichum gloeosporioides* Penz. which is conidial stage of *Glomerella cingulata* (Ston.) Spauld and Shrenk (Small, 1924 and 1926). Three species of *Gloeosporium* (*G. mangae* Noack, *G. mangiferae* P. Henn and *G. raciborskii* P. Henn.) have been described on mango by Saccardo (1902). *Colletotrichum gloeosporioides* var. *minor* was reported on mango and papaya in Queensland by Simmonds (1965). Fitzell (1979) reported *C. acutatum* causing anthracnose of mango in New South Wales.

Physiology of the causal organism

The fungus produces good aerial mycelium in Richard's and Browns's agar and profusely sporulates on oat meal and corn meal agar along with abundant development of acervuli in rings and few setae. Optimum pH lies between 5.8 and 6.5. The fungus grows at 15-35°C with an optimum at 25°C. Growth ceases beyond 35°C. Glutamic acid and alanine supported best growth and sporulation. Of the various monosaccharides tried galactose supported best growth followed by glucose and xylose. Fungus failed to utilize galactose upto 15 days while glucose was depleted from the medium within 6 days (Tandon, 1967). Trehalose produced best growth on 10th day while melezitose produced maximum growth at 15th day. Sucrose was not detected in the medium after 5 days while melezitose was present upto 10th day.

Post infection changes

There was no rot when fruits inoculated with *C. gloeosporioides* were incubated at 15°C for 10 days but some decay was observed after 15 days. It was observed that 'Totapari' variety could tolerate lower temperature better than 'Langra. Soluble sugars detected in the mango fruit were sucrose, glucose, fructose and monosaccharide (Rf 0.75). After 4-6 days of infection oligosaccharide (Rf 0.27) appears. At this stage a rapid decrease in the sucrose content indicated that sucrose contributed to the formation of this oligosaccharide. This was confirmed by growing the pathogen in an artificial liquid medium containing sucrose where a synthetic oligosaccharide (Rf 0.27) was also formed.

Loss in Vit. C content was more in 'Dashehri' as compared to 'Langra' (Tandon, 1967). After 4 days incubation, Ghosh *et al.* (1966) obtained substantial decrease in Vit. C.

Control

Diseased twigs and leaves which fall on the ground are a potential source of infection. Such diseased material should be collected and all infected twigs from the tree should be pruned away and such refuse should be burnt.

Sanitation and application of protective sprays and development of resistant varieties has been practiced in Philippines (Ocfemia and Agati, 1925). Sattar and Malik (1939) recommended spraying of young plants in the nursery with Bordeaux mixture (3:3:50). Hendrix *et al.* (1957) tested Dichlone, Ziram, Ferbam and other organic fungicides on several mango varieties. Vender Weyen (1964) reported Zineb, Captan and Bordeaux mixture on panicles in Morocco. Gadre (1979, 1982) achieved good control of anthracnose disease by spraying of fungicides like Bordeaux mixture, Captan, Blitox and Difolatan. McMillan (1971) reported that Benomyl was effective, when compared with Maneb followed by tribasic copper sulphate and Captafol. Wettable sulphur 0.2% before flowering Carbendazim 0.3% at pea stage Zineb 0.2% before maturation of stone reduced the incidence of the anthracnose (Jadeja and Vaishnav, 1984). Use of Zineb and Bordeaux mixture (4:4:50) is recommended by Tandon and Singh (1968).

Since stem, leaves, flower panicles and fruits are affected, tree sanitation is very imporatant to minimize the disease infestation. Foliar infection in the hot humid parts of West Bengal may easily be controlled by spraying of Captaf (2 g/litre) or Bavistin (1 g/litre) or Topsin M-70wp (1 g/litre) 3-4 times during January—April (Dhua and Raychaudhuri, 1996). Pre-harvest spraying of systemic fungicides Topsin or Bavistin twice or thrice at 15 days intervals commencing from 50-55 days of fruit set are quite effective to control losses during transit and storage. Elliot and Patterson (2000) evaluated effectiveness of selected fungicides by Poison agar test against *Colletotrichum gloeosporioides*. The fungicide concentrations ranged from 0 to 0.1%. Benomyl, Carbendazim, Thiophanate-methyl, Copper hydroxide, Sulphur and Tribasic copper sulphate. Diathanea (a dithiocarbamate) and Chlorothalonil were selected for field trials.

Chung (2001) studied sensitivity of mango anthracnose pathogen to fungicide prochloraz in Taiwan.

To control the post-harvest fruit rot, hot water treatment was also tried. Smoot and Segall (1963) recommended dipping at 54.5-55.5°C for 5 min. Tandon and Singh (1968) found 50-55°C effective for 15 minutes. 55°C for 5 minutes, added with Benomyl or TBZ (0.1%) (Spalding and Reader, 1972) gave good control of anthracnose. Treatments of fruits with heated Benomyl at 55°C for 5 min yields satisfactory results (Jacobs *et al.*, 1973).

1.14. SCAB

The mango scab was discovered from Florida by Ruchle (1943). The scab fungus *Elsinoc mangiferae* Bitan & Jen (Bitan-court and Jenkins, 1946) attacks leaves, twigs,

panicles, blossoms and causes blotches on the bark of the stem and spots on the mango fruit.

The lesions differ in size and colour depending upon age of the plant and the country in which the disease occurs. These spots in Cuban nursery, are pale to brown and covered with delicate buff down, which represent the vegetative stage of fungus but on older foliage, the lesions are bigger in size and are grey in appearance surrounded by dark margin and bear on their surface small dot like structure, which are ascomata of the *Elsinoc*. Symptoms produced by the disease are very much like those of anthracnose. Severe attacks cause crinkling and distortion of the leaf, followed by premature shedding.

Control

Frequent sprays of copper fungicides to protect new flushes of growth are effective for scab control in nurseries. Wettable cuprous oxide (1 lb in 100 gals water) or other neutral copper fungicides have proved useful (Ruchle and Ledin, 1955).

1.15. SOOTY MOULD

Sooty mould or sooty blotch on mango is common wherever honey dew or sugary substance secreting insects *viz.* mango hoppers, scales, coccids and mealy bugs are found. The disease has been reported from Hawaii, Mauritius, South Africa, Formosa, Pakistan and India (Prakash and Srivastava, 1987).

It is a common malady of fruit orchards in India. The fungi are not parasitic because host pathogen relation is not observed in this case. The disease spreads through insects namely *Lecanium hemispharicum, L. nirum, Hemilecanium inbricans* etc. Sooty moulds grow rapidly in exudates secreted by these insects and injuries on twigs and leaves created by above insects. Fungi produce numerous black spores which are stuck on secretion. Due to undesired black growth the leaves become discoloured and unable to photosynthesize. Often young plants are perished completely due to heavy infection. Affected leaves curl and shrivel under dry conditions. Disease is severe in old and dense orchards where light intensity is low.

Sooty mould is caused by the fungus *Meliola mangiferae* Earle (Butler and Bisby, 1931 ; Uppal, 1935), *Capnodium mangiferae* Cke and Brown (Vaheeduddin 1953) and *C. ramosum* Cke (Butler and Bisby 1931). Das and Mohanty (1972) have reported that sooty mould can be incited by *Tripospermum acerinum* (Syd.) Speg. White shooty blotch of mango is caused by *Gloeodes pomigena* (Schw.) Colby (Pole Evans, 1934).

Control

The remedy of this disease consists in killing the insects by suitable insecticides. The mould will die out for want of suitable growth medium. Singh and Singh (1972) reported the spraying of Elosal (900 g/450 *l* of water) at 10-15 days intervals to be quite effective. The fungal growth can be checked by applying Wettable Sulphur (0.2%).

If the incidence is noticed on the fruit, dipping the fruit in bleaching solution (for 2 min. in 1/4 lb each of chloride of lime and boric acid/gallon of water) and subsequent washing in clean water is recommended (Wager, 1937).

1.16. PINK DISEASE

The disease was reported from South India by Subba Rao (1936) and Mc Dougall (1940). It is known as thread blight, rubellosis and cofweb. The disease is prevalent in many tropical and subtropical countries like India, Sri Lanka and Brazil.

First symptoms are pinkish powdery coating on twigs and branches. Fungus invade inside tissue and girdle the stem. It interfers with the transport of nutrients. Severely infected bark becomes shredded and wood is exposed, leaves become yellow and dry, shoot and branches of tree wilt and ultimately dry. The pink colour on the tissues represent profuse conidial production by the fungus and hence the name 'Pink Disease'.

Deslandes (1944) reported thread blight which causes extensive shrivelling and whitening of the crown, in regions with dense and heavily shaded plantings. The mid lateral veins of leaves are coated heavily with the fungal mycelium.

The disease is caused by *Botryobasidium salmonicolor* (Berk. and Br.) Venkatrayan

(Syn. *Pellicularia salmonicolor* (Berk, & Br.) or *Corticium salmonicolor* Berk and Br.) Another species *Corticium koleraga* (Cke.) V. Hohn causes thread blight in Brazil.

Control

1. The disease can be controlled by cutting and burning of infected twigs.
2. Bordeaux mixture should be sprayed and Bordeaux paste should beapplied to cut ends.
3. Use lime sulphur and oil based coppers.

1.17. BLACK BANDED

The disease was recorded by Massee from Pune (Saccardo, 1906). It has been observed in several states including Gujarat (Valve *et al.*, 1985). The disease is caused by *Rhninocladium corticolum* Massee which is an imperfect stage of *Peziotrichum corticolum* (Massee) (Subramanian, 1956).

Black velvety fungal growth covers the major part of the lateral branches and twigs. Infection occurs in midrib and veins of leaves. The incidence of the disease is very low on the main branches. It presents a characteristic and conspicuous black banded appearance. The infected portions of the bark usually show the mycelial growth and cluster of conidiophores, thereby presenting a velvety appearance during the rainy season. The mycelial growth drops off in the summer months leaving light black bands in the affected portions. The fungus is confined to the upper layer of bark (Reddy *et al.*, 1961).

Control

The application of Bordeaux paste or Bordeaux mixture may control the disease followed by gunny rubbing or Ganji treatment (Reddy *et al.*, 1961).

1.18. CERCOSPORA LEAF SPOT

The disease is caused by *Stigmina* (*Cercospora*) *mangiferae* (Koorders) Ellis and *C. mangiferae indicae* Munjal, Lal and Chona. The leaf spot disease is reported from several countries.

The disease is characterized by the presence of irregular or round lesions on the both sides of the leaf blade, measuring 3.5—10 mm in diameter. Initially the lesions are light brown but soon after they become brown to black in colour. The lesions are surrounded by yellow halo and center somewhat dull in colour. A few to over a hundred spots may be seen on a leaf blade, depending upon the severity of infection. Several spots may coalesce to form large irregular patches. The affected portions of the leaf blade may soon dry off and wither, resulting in a scorched appearance of the foliage.

Control

Spray of copper fungicide and oil is recommended by Frossard (1959).

1.19. ALTERNARIA SPOTS

The disease is noticed on leaf, twigs and fruits. Mukherji and Bhattacharya (1965) reported the disease caused by *Alternaria tenuissima* (Fr.) Wiltsch. Yadav and Udainarain (1970) reported *A. alternata* (Fr.) Keissler, (*A. tenuis*) from the fallen stem of mango. Srivastave *et al.* (1964) isolated it from the fruit.

Symptoms appear as small, brownish circular spots on the surface of leaves and fruits and as black patches on the twigs. Initially it became apparent on lower surface of the leaf where it shows light brown colouration. After a few days the infection is visible on the upper surface of the leaf also. The spots gradually enlarge and become irregular black and form larger patches. The younger leaves are found to be more susceptible than mature ones.

On fruit, the spots enlarge subsequently becoming water-soaked and in advanced stages the whole fruit becomes rotten. Below the skin, reddish brown patches can be seen on the flesh. Pandey *et al.* (1983) reported premature fruit fall by *A. alternata*. It starts with drying and shrivelling of the stalk end which changes to brown from normal green colour. The immature fruit falls down due to abscission layer formation. Artificial inoculation on Dashehri and Safeda Lucknow resulted in 10-25% loss. Singh and Tandon (1967) found that isolates of *A. tenuis* from banana, guava and mango were pathogenic under artificial conditions except the isolate from papaya. Leaf isolates failed to infect the fruits.

Control

The control of *Alternaria* leaf spots can be achieved by application of Cuprosan super D, followed by Bordeaux mixture. Mukherji and Bhattacharya (1965) used Merculin followed by mercurized copper oxychloride. Antibiotic Aureofungin was found effective by Sinha and Bilgrami (1987).

1.20. PESTALOTIA ROT

The disease is known as Pestalotia leaf spot or Grey blight. It also affects the green and mature fruit :

Pathogens

Pestalotiopsis mangiferae (P.Henn.) Steyaert, Prakash and Srivastava, (1987).

Pestalotia sp. (*Calonectria mangiferae*) Sechet (1955).

P. dichaeta Speg., Viegas (1946).

P. funerea, Wardlaw and Leonard (1936).

P. guepini, Steyaert (1953).

P. estalotiopsis sp., Anon (1974-1985).

P. versicolor (Speg.) Stey. Anon, (1974-1985).

P. funerea var. *mangiferae* Uppal *et al.* (1935).

P. glandicola (Cast) Stey., Ullasa and Rawel (1985).

The disease is characterized by the presence of brown spots on the lamina of mango leaves, more commonly during the winter season. Initially the spots are brown and minute which gradually increase in size and become dark in colour. Some of the spots often enlarge considerably to form large lesions, in which the central region turns greyish white or light olive grey in colour with an encircling ring of tan colour. At this stage black spots of acervuli may become visible to the naked eye in the central region, more on upper surface of the leaf, and never extend beyond the midrib of the leaf, but if the infection starts from the tip it advances regularly on either side of the mid- rib. Finally, the infected portion gets detached from the leaf after 3-4 months. In cases of severe infection defoliation occurs.

Appearance of the infection on mature green fruit is characterized by the presence of the brown small spots which gradually increase in size, become dark coloured with greyish white centre having black pin-head like dots of acervuli. Such a fruit drops off when it advances to the stalk-end, wounding results in quick attack by the fungus and early spreading of lesion. Moist atmosphere helps in increasing and hastening infection (Sarkar, 1960).

Symptoms on a mature fruit include dark brown, circular lesions extending all over the fruit, at advanced stage it assumes dark blackish colouration and sticky fluid comes out very common on the ripe fruit. Sometimes hyphal growth with black pustules covers the infected part.

When *P. mangiferae* was artificially inoculated to different varieties of the mango fruit, it was found to cause severe rotting of the fruits of 'Langra', Dashehri, Chausa, Banglora and Fazli varieties (Tandon, 1967).

Physiological Studies

Fungus grew well in Potato Dextrose Agar, Czapek's Dox Agar and Richards Agar media (Sarkar, 1960). Pandey and Mohammad (1974-75) obtained maximum growth and sporulation in Host Extract followed by Richard's media. The fungus was capable of growing at temperatures between 10° and 35°C and the optimum lies between 20-25°C. Best mycelial growth and excellent sporulation takes place between pH 5.5-6. The temperature range for spore germination is 10-34°C with an optimum at 30°C.

Control

1. Dusting of leaves with zinc sulphate controlled the disease but similardusting on the fruit failed to control the rot (Tandon *et al.*, 1955).
2. Jadeja and Vaishnav (1984) used Wettable Sulphur (0.2%) + Zineb (0.2%) after completion of heavy showers followed by Wettable Sulphur (0.2%) before flowering, Carbendazim (0.3%) at pea stage and Zineb (0.2%) before maturation of stone which reduced the incidence.
3. Of the five antibiotics tried Aureofungin had greatest inhibitory effect (Sinha and Bilgrami, 1987).

4. Efficacy of homeopathic drugs was tested in controlling the fruit rot of mango. Khanna and Chandra (1978) found lycopodium clavatum potency 190 effective in reducing the post-harvest fruit rot. Out of the five adjuvents tested by Khanna and Chandra (1989) soap solution was most effective. Their studies showed that Homeopathic drugs not only inhibit spore germination and growth of fungi *in vitro* but are effective also in the infection court.

1.21. PHOMA LEAF SPOT AND FRUIT ROT

Phoma sorghina (Sacc.) Boerema, Doren. and Vankest causes leaf spot of mango (Prakash and Singh, 1976 ; Prakash and Raoof, 1985). Arya and Mathew (1991) found *Phoma multirostrata* (Mathur *et al.*) Dorenbosch and Boerema affecting ripe mangoes. The disease manifests itself in the form of small irregular, oval to roughly circular water soaked spots on young leaves measuring from pin head to 2.5 mm in size. Lesions are brown later differentiated into brown margin with straw colour. Yellow halo around the brown margin is also observed. The infected leaves become brown and ultimately dry up. Symptoms produced by this fungus are very much similar to anthracnose. However, in this case spots are smaller and there is no cracking in the centre as found in anthracnose.

P. multirostrata produced characteristic sunken brown coloured spots which gradually covered the whole fruit. Numerous small pin-head like fruiting bodies (pycnidia) were observed on the surface of the fruit in advanced stages. Fruits of papaya, litchi, banana, guava, plum and apple were found susceptible to this organism.

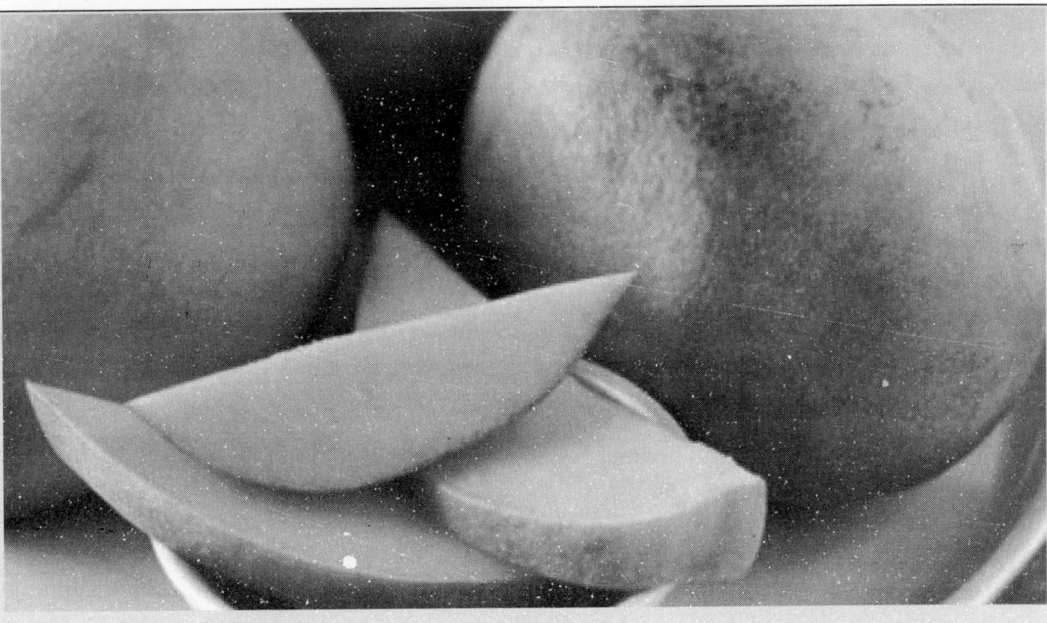

(A)

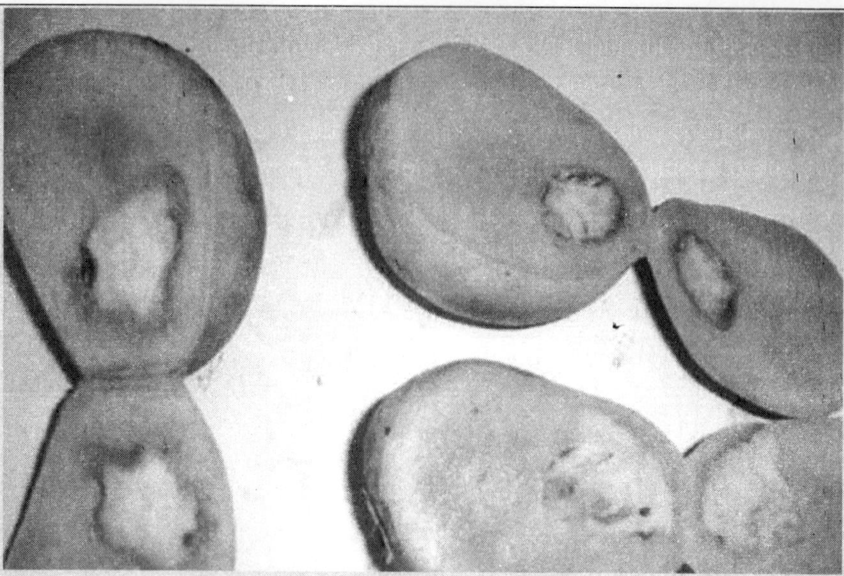

(B)

Fig. 12. (A) Healthy (B) Infected fruit due to *Phoma multirostrata.*

Control

Three systemic fungicides tried *viz.* Bavistin, Derosol and Delan were significantly effective in controlling the disease. However, Derosol was most effective controlling 95 per cent disease at 750 ppm concentration (Arya and Mathew, 1991).

1.22. LOPHODERMIUM LEAF SPOT

The disease is caused by *Lophodermium mangiferae* Koord. It was reported from Varanasi by Tiwari and Srivastava (1977).

The fungus produces amphigenous, whitish, scattered, irregular, 4-6 mm diameter spots with black speck like ascocarps in the centre. A vertical section through the apothecium shows the paraphyses and asci within. The individual asci are cylindrical containing 8 filiform, multiguttulate ascospores.

1.23. CURVULARIA LEAF SPOT

The disease was first reported from Allahabad by Singh (1971). It is caused by *Curvularia lunata* (Wakk.) Boed.

Light grey coloured spots are formed on the tip of the leaf. These spots proceed towards the petiole due to which entire leaf surface becomes dark grey. Fruit spots were also noticed.

1.24. ASPERGILLUS ROT

Aspergillus niger van Tiegh. was found responsible for Aspergillus rot in various parts of the country (Srivastava *et al.*, 1965). The pathogen caused 20-22% loss in Totapari and

Langra varieties in Allahabad in June 1964 and 35% loss in Safeda Lucknow variety in Lucknow in July, 1964. The disease is more serious during storage and marketing processes.

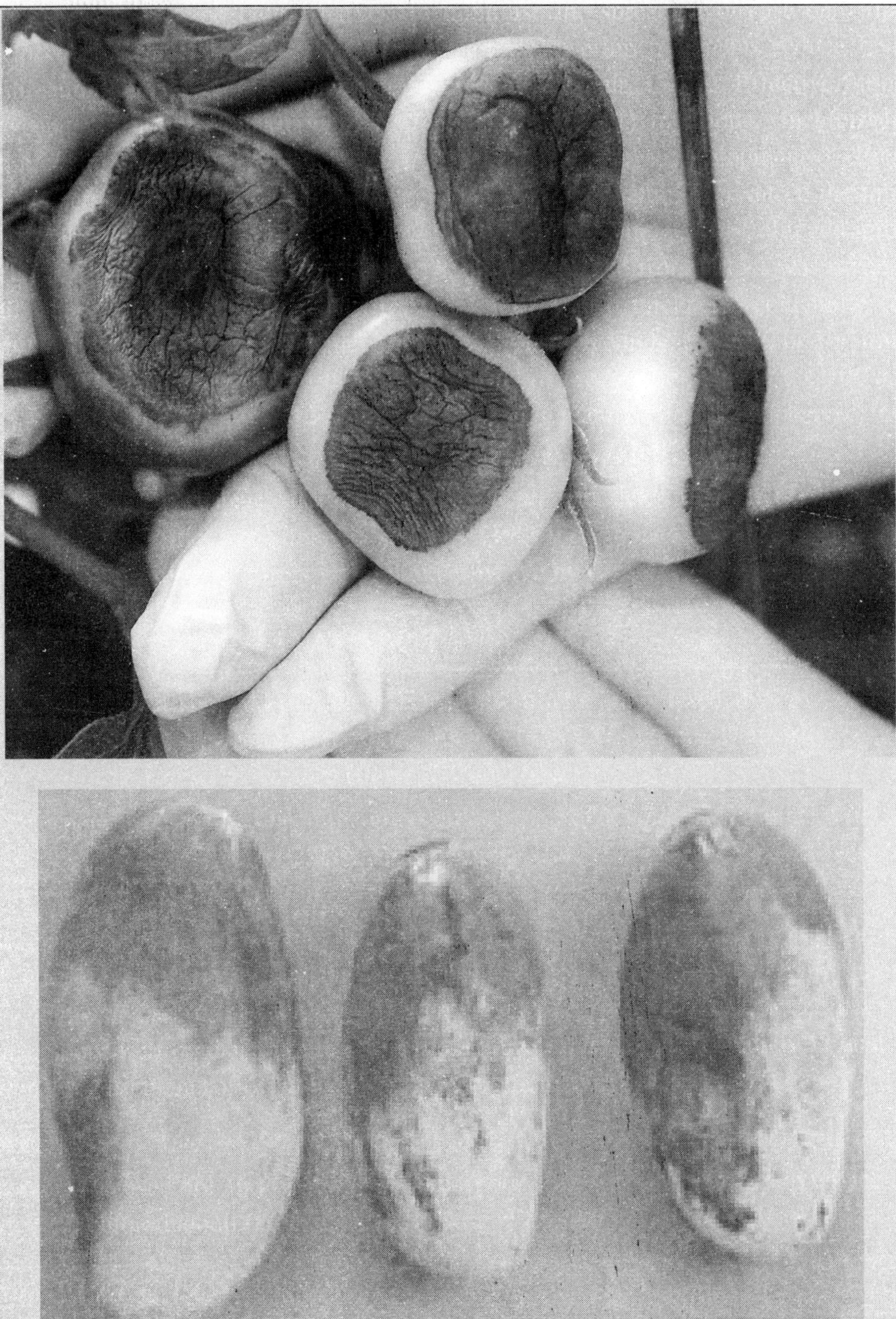

Fig. 13. Stalk end rot due to Aspergillus.

Fig. 14. Aspergillus rot.

The disease usually appears at the stalk end of the fruit and is quite distinct from rot caused by other organisms. At first it appears as a light brown circular patch. The intensity of colour of infected area depends upon the variety of the fruit and the stage of the disease. The lesion grows in a regular manner and forms a large circular spot encircling the stalk. After three or four days the infected region becomes sunken and at an advance stage it becomes somewhat crumpled. The rot is soft in nature. Black conidial heads of the causal organism become distinct after five or six days.

Besides, causing stalk-end rot *A. niger* also causes lateral rot, tip-end rot or a rot at any other place. In these cases the position of the lesions depends on the position of injury on the fruits. Sometimes the spots are irregular in shape and this might be due to different types of injury caused to the fruit.

A. niger can grow both saprophytically as well as parasitically. Srivastava *et al.* (1965) isolated it from soils as well as dead or decayed twigs or leaves of different varieties of the mango. The organism has been isolated from Dashehari, Safeda Lucknow, Makhan, Taimurity and Langra from Lucknow, Allahabad and Varanasi (U.P.).

This organism has also been found to cause damage to the fruit in plantation. Krishan Bhogbela variety of the mango was severely damaged by *A. niger* in the orchards in Krishnagar (West Bengal). Besides being badly rotton a number of fruits had become mummified on account of loss of water.

Such a fruit remained attached to the plants and were thus capable of supplying inoculum to the other developing fruits. It is thus clear that either Aspergillus rot appears quite early and damages the fruit in plantations or the pathogen remains dormant and becomes responsible or post-harvest decay of the fruit at a later stage.

Control

(1) Bhargava and Singh (1975) found 1500 ppm Benlate effective against the disease, similar to Jacobs *et al.* (1973). Tecto 40 (a formulation of TBZ) was effective at

4000 ppm but it deteriorated the flavour and taste of the fruit hence not recommended.

(2) Saprol and Delan could control the rot at 1000 and 1250 ppm respectively (Pandey *et al.*, 1980).

(3) Hot water treatment (55°C for 5 min) delayed the rot for 6 days, (Pathak and Sekhawat, 1976).

1.25. RHIZOPUS ROT

Thakur and Chenulu (1966) reported *Rhizopus arrhizus* alone, responsible for .3 per cent decay in mango fruit. Thakur (1972) found two more species of *Rhizopus viz., R. oryzae* and *R. stolonifer* causing soft rot in apple, banana, mango, potato and tomato.

Disease appears as light brown coloured circular patches. The intensity of the colour varies according to the variety and duration of time. The lesions grow in regular manner and forms large circular patches. The wall of the fruit becomes soft. In severe cases watery ooze comes out. Effect of temperature and humidity n disease development was studied by Thakur (1972). He found a wide range of temperature (7—40°C) effective against the disease. At low temperature (7—8°C) the disease development was slow and of low intensity, whereas, at temperature between 20-40°C the disease development was very rapid, 100% fruit rotted in 4 days at 10- 25°C while the rot was 100% in 2 days at 30 to 40°C. A relative humidity of 50 per cent and above allowed rapid development of soft rot of moderate to heavy intensity.

Control

(1) Pre-inoculation treatment with Thiabendazole completely controlled *R.arrhizus* upto 20 days.

(2) Combined action of 2-aminothiazole with 2-aminopyridine (each in 5per cent concentration) protected mango fruits for 20 days (Thakur and Chenelu, 1970). After 20 days, however, the rotting was rapid and all the chemicals failed to p rotect the fruits from rotting.

1.26. PHOMOPSIS ROT

The disease is caused by *Phomopsis amraii*, Srivastava *et al.* (Tandon *et al.,* 1965) and *P. mangiferae* Ahmad (Laxminarayana and Reddy, 1975 and Reddy and Laxminarayana, 1978). This rot starts from the stalk end, dark brown, slightly irregular in outline, soft but not watery, at an advanced stage the fruit got detached from the stalk and dropped. The disease was observed commonly on Langra, Chausa, Bombai, Dashehari, Safeda, Rataul, Tamuria and Himsagar varieties (Tandon, 1967).

Control

Use of Bavistin and Benlate at 1000 ppm can control the rot.

1.27. LASIODIPLODIA (BOTRYODIPLODIA) ROT

Botryodiplodia theobromae Pat. (Syn. *Lasiodiplodia theobromae*) causes soft rot of mango (Srivastava *et al.*, 1965 ; Tandon, 1967). The same pathogen is reported to cause Die-back diseases in a number of trees (See 1.11.). The disease causes considerable loss during storage, marketing and transportation.

The infection usually starts at the stalk end of the fruit and very rarely at other regions. It begins as a water-soaked lesion surrounding the stalk which increases and soon covers up about 1/3rd to 2/3rd of the fruit. At this stage the central part of the infected area acquires a dark brown colouration and is surrounded by a light brown peripheral zone, the irregular margin of which assumes a water-soaked appearance. Pycnidia develop at a later stage of rot as minute bodies. The rot is soft in nature.

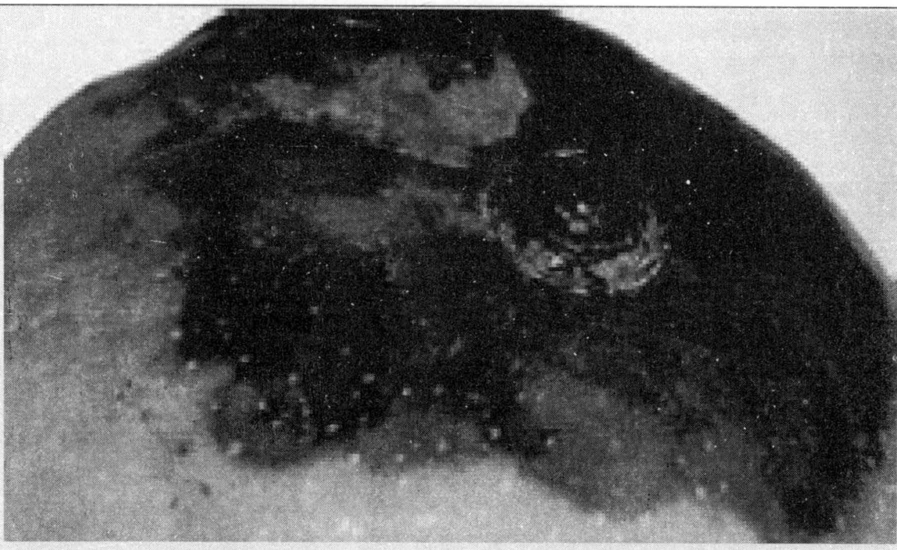

Fig. 15. Stalk rot of mango caused by *Botryodiplodia theobromae*.

The association of this fungi with other parts of mango plants suggests that the organisms may get associated with the fruit in the orchards and subsequently cause infection when the fruit is kept in storage. It has been observed that mostly the injured fruit becomes exposed to the possibilities of getting infected with this pathogen.

Physiological studies were carried out by Tandon (1967). Mango isolate of *B. theobromae* could grow between 15-35°C with an optimum of 25°C. It grew between 2.5 to 10.5 with an optimum at 6.0. Maximum sporulation and dry weight was achieved on mannose, followed by sorbose and arabinose. Of the 8 monosaccharides used arabinose was utilized in 5 days followed by glucose and rhamnose 9 days and mannose in 12 days. Of the oligosaccharides, cellobiose supported best growth. L-asparagine supported best growth of *B. theobromae*.

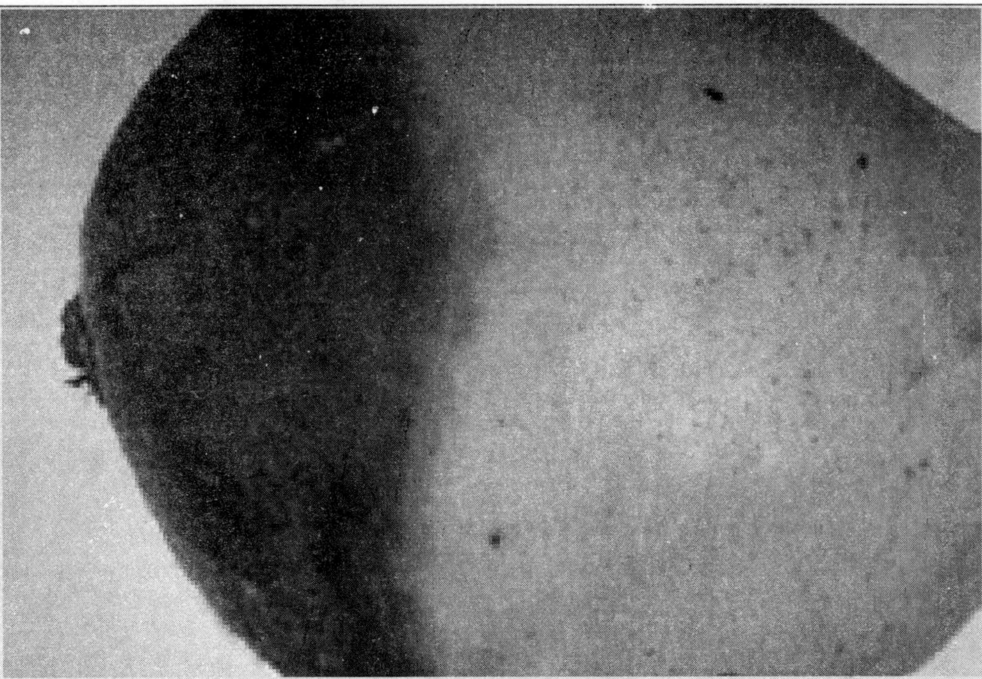

Fig. 16. Stalk end rot due to *Lasiodiplodia* sp.

Physiological and post infectional studies revealed that this organism caused considerable decrease in the amount of sucrose. An increase in glucose and fructose become evident after 4 and 8 days of incubation. At the end of the incubation period of 12 days, sucrose and an unknown monosaccharide (Rf 0.72) were not traceable while glucose and fructose were present in small quantities. The fruit in'fected with *Botryodiplodia* showed gradual decrease in vitamin C content but this decline was more rapid as compared to the healthy fruit.

Control

(1) Use of Bavistin is reported to control the disease.

(2) Pandey *et al.* (1983) reported that mustard and castor oil checked the rot completely at 75% concentration as pre- and post-inoculation dip. Linseed and Mahua oils could check 96.0 and 95.85 per cent as pre-inoculation treatment. Maximum prevention was upto 97.2% and 95.7% in linseed and Mahua oil respectively. Mahua oil developed slightly unpleasant smell hence not recommended. When wrapping with oil paper was done 100% control was achieved in Castor and Mustard oil impregnated paper.

(3) Meah, (2000) studied retention of residues of Mancozeb and Iprodone in mango fruits treated for control of stem end rot in Pakistan.

(4) Bambosa *et al.* (2000) studied effect of ozone, iodine and chlorine on post-harvest diseases of mango caused by *Colletotrichum gloeosporioides, Fusarium oxysporum, Pestalotiopsis mangiferae* and *Lasiodiplodia theobromae*. Earlier chlorine has been used for disease suppression due to its oxidizing properties.

1.28. MEMBERS OF BASIDIOMYCETES ON MANGO

Various fungi are reported from the old stem, branches and trunk of the mango tree, which belong to the Basidiomycetes group, but it is not clear whether, these fungi may infect the plant or only saprophytically grow upon the older part of the tree. If these are pathogenic, the infection is local. The leaves become discoloured and fall, gummosis takes place. Following fungi are reported which belong to Basidiomycetes on the mango tree from India :

(i) *Fomes conchatus* (Pers.) Fr. This fungus is reported from Bengal and Maharashtra (Bose, 1934, Banerjee, 1947).

(ii) *Ganoderma appalnatum* (Pers.) Pat. It is reported from Tamilnadu and Kerala (Banerjee, 1947). Due to the infection caused by this fungus dry rot of the stem takes place and finally the stem dies. It may take many years.

(iii) *Polyporus gilvus* Schw., and *P. tulipferae* (Schw.) Overholts are reported from North India (Bakshi *et al.* 1956). The sap wood of the tree starts rotting and it becomes spongy.

(iv) *Schizophyllum commune* Fr. This fungus is reported from Punjab (Anon, 1950).

1.29. TABLE. SOME MORE FUNGI ISOLATED FROM DIFFERENT PARTS OF MANGO TREE

Parts used for Isolation	Organism	Locality	Reference
1	2	3	4
Seedling Leaf	*Chlamydomyces sp.* *Nigrospora oryzae* (Berk, et Br) Petch *Pithomyces saccharii* (Speg.) Ellis *Helminthosporium compactum Karst* *Phomopsis amraii*	Lucknow Cape Camorin Cape Camorin Rataul (Meerut) Allahabad Bombay Jodhpur Egypt Allahabad & Lucknow	Tandon, 1967 Tandon, 1967 Tandon, 1967 Srivastava *et al.* 1964a. Tandon, 1965 Rao *et al.* 1977. Reddy, 1968 & Abo-el-Dahab, 1966 Srivastava, *et al.*, 1964a.
	Srivastava *et al.* *Cytospora mangiferae* Sharma *Aureobasidium pullulanse* (deBary) Arnand. *Stemphyllium Wallr. sp.* *Phyllosticta mortoni* Fair.		

Stem and twig blight	*Diplodia natalensis* Evans *Nigrospora sphaerica* (Sacc.) Mason *Chaetomium sp. Gloeosporium sp. Fusarium* sp. *Phyllosticta mortoni* Fair.	Varanasi, Allahabad, Lucknow & Meerut Sabour Sabour Patna Rataul (Meerut) Allahabad and Lucknow	Srivastava, *et al.*, 1964a. Tandon, 1967. Tandon, 1967. Tandon, 1967. Srivastava, *et al.*, 1964a. Srivastava, *et al.*, 1964a.
Flowers and Young fruit	*Chlamydomyces sp. Cladosporiu herbarum* (Pers.) Link *Curvularia lunata* (Wakk.) Boedj *Fusarium sambucinum* Fuckel *Melanospora damnosa* (Sacc.) lind. *Rhizopus nigricans* Ehrenb. *Malustela aeria* Batista, Lima and Vasconcelos *Oospora* sp. *Bartalinia* sp.	Rataul (Meerut) Malihabad (Lucknow) Meerut Meerut Meerut & Lucknow Lucknow Lucknow Coimbatore Lucknow	Srivastava, *et al.*, 1964a. Srivastava, *et al.*, 1964a. Srivastava, *et al.*, 1964a. Srivastava, *et al.*, 1964a. Tandon, 1967 Tandon, 1967 Tandon, 1967 Tandon, 1967.
Fruit rot	*Cochliobolus spicifer* Nelson *Helminthosporium hawaiiense* Bougnicourt. *Macrophomina phaseoli* (Maubl.) Ashby *Helminthosporium compactum* Karst. *Mucor* sp.	Meerut Delhi Patna Allahabad Allahabad	Tandon, 1965 Tandon, 1965 Tandon 1965 Tandon, 1967Tandon 1967.

DISEASES CAUSED BY BACTERIA

1.30. BACTERIAL CANKER

Canker disease is one of the most common disease of mango in many countries (Prakash and Srivastava, 1987). It is also known as Bacterial spot, Leaf spot, Black spot, Mango blight and Bacterial black spot.

It is reported from almost all mango growing states. Herbarium specimens collected as far back as 1881 have been found to show the presence of disease spots.

Doidge (1915) identified the pathogen as *Bacillus mangiferae* Doidge but this name was later changed to *Erwinia mangiferae* (Breed, Murray and Smith, 1957). In India the disease is caused by *Pseudomonas mangiferae indicae* (Patel *et al.*, 1948 a, b).

The disease attacks leaves, leaf stalk, twigs, branches, inflorescence and fruits. The severity of the disease is noticeable in Lucknow region after June-July, sometimes in May, initially water soaked lesions are formed which turn into typical cankers. On leaves the disease first appears as minute water soaked irregular stellate to angular raised lesions 1-4 mm in diameter, usually more on the apex. The lesions are light yellow in colour with yellow halo but with age enlarge or coalesce to form irregular necrotic cankerous patches. This usually occurs under ideal conditions for infection. In severe cases the leaves turn yellow and drop down.

On branches, twigs and stem the fresh lesions are water soaked, swollen, which later become raised and dark brown with longitudinal fissures exposing the vascular tissues mostly filled with gum which oozes out. The flower stalks are often attacked resulting in a serious fall of flowers and young fruit (Doidage, 1932). When humid conditions prevail at the time of flowering and fruit setting, serious losses may occur as a result of the attack of pathogen and at this proper juncture, the bacterial disease resembles anthracnose (Doidge, 1932).

On fruits the initial water soaked lesions gradually develop into cankers. The dark centre grows bigger usually upto 5 mm in diameter, which sometimes run almost the whole length of the fruit. They often burst open, releasing a gummy ooze containing bacteria. The canker exposes the flesh which attracts insects and in many cases becomes invaded by secondary microorganism which causes rotting. Fruit dropping is more when infection is near stalk end. Unhealthy mangoes are unfit for sale.

Control

(1) Removal of infected material, orchard sanitation, seedling certification and disinfection of pruning knife are recommended.

(2) Disease can be controlled by Bordeaux mixture (4:4:50) at the rate of 7.5 gal per tree (Wager, 1937). Bordrick (1971) recommended Copper oxychloride or Cuprousoxide for the control of disease. Spray of Copper containing materials and Agrimycin was recommended by Viljoen and Kotze (1972). Injecting Bavistin (1000 ppm) followed by Plantomycin (Ramakishun, 1981, 1985) and Streptomycin sulphate (0.025%) followed by Aureofungin (Prakash and Raoof, 1985) have been found effective in controlling the bacterial canker of mango.

1.31. BACILLUS SOFT ROT

Patel *et al.* (1950) reported *Bacillus subtilis* causing soft rot of mango. Earlier Patel and Padhye (1948) have reported *Bacterium carotovorum.* They conducted pathogenicity tests and found mango, banana, green and ripe tomato positive while infection was not produced on cucumber, carrot, lady's finger, French beans, cowpea, raddish, brinjal, onion, garlic, apple, lemon, *Citrus sinensis* etc.

The bacteria is a gram positive, non-capsulated rod with rounded ends. They are motile with peritrichous flagella, creamy white, not acid fast, spore formers, single or in chains and stain readily with common dyes much as methylene blue and Carbol fuchsin.

DISEASE CAUSED BY ALGA

1.32. RED RUST

The disease is caused by an alga *Cephaleuros virescens* Kunze (Syn. *C. arasiticus* Karst ; *C. mycoidea* Karst). It is limited to all continents and probably all islands between 32°N and 32°S provided the temperature and humidity are suitable for their growth and reproduction. The disease appeared in an epidemic form in the Tarai region of Uttranchal in 1956 and was reported to cause reduction in photosynthetic activity and defoliation, resulting in lowering the vitality of the plant (Pathak, 1980).

The symptoms on the host is characteristic and are easily recognised by the presence of rusty red fructification of the alga on the surface of the infected leaves, veins, petioles and young twigs. Initially the spots are greenish grey in colour and velvety in texture and finally they turn reddish brown. Spots are circular to irreglar in shape, erumpent (raised) measuring 2 mm in diameter, when several spots join it may be 1 cm. After shedding of spores the algal matrix remains attached to the leaf surface, leaving a creamy white mark at the original rust spot (Singh, 1960).

The alga thrives on the leaves, bark and twigs of the host plant as parasite and epiphyte and causes bark lesions on young branches and stem blight on grown up trees. The alga is generally shed off by exfoliation of outer tissues when plants are vigorous. The disease is rare on newly emerging shoots. Fruit infection is encountered infrequently. Many dark brown blemishes are produced on the fruits.

Foliar spots vary from mere specks to big patches. Alga extends at first between the cuticle and epidermis and then penetrates the epidermal cells. Penetration of the leaf blade is complete and affected cells eventually die.

The physiology of alga in infected leaf tissue was studied (Vidhyasekaran and Parambaramani, 1971a, b) Glucose and sucrose were less, fructose was more. Starch, cellulose and pectin were found to be more while lignin was found to be unaffected. The dry weight of leaves increased due to infection. Disease caused reduction in the total protein, ammoniacal nitrate, amino and amide nitrogen content of mango leaves, whereas nitrates accumulated in the infected leaves. Glutamic acid and alanine increased markedly while glycine decreased sharply whereas valine content was not affected by the algal infection. Vidhyasekaran and Parambaramani (1972) observed an increase in transpiration and decrease in water content while K and P content was less in the infected tissues. Na accumulated in them, Mg content slightly increased while S content decreased due to infection. Among the trace elements, iron and manganese accumulated while copper decreased in diseased tissues.

Control

1. The disease can be reduced by planting the trees at proper distances.
2. Pruning and manuring of the trees is beneficial.
3. Regular sprays with copper fungicides (Ruehle and Ledin, 1955), or Copper sulphate (Suit and Ducharme, 1946) is effective.
4. Prakash and Singh (1979) recommended Bordeaux mixture (5:5:50) followed by blue copper to control the disease.
5. Spary of fungicides like Difolatan followed by Bordeaux mixture is suggested by Gupta *et al.* (1980).

PHYSIOLOGICAL DISORDERS

1.33. BLACK TIP

In U.P. the disease is known as Koeli, its other names are Taper tip, Girdlenecrosis, Koel Padda and Chimney disease. It was first reported from Bihar in the year 1908 (Woodhouse, 1909). In U.P. it was noticed in 1923, since then it has been reported to occur in West Bengal, Punjab, Haryana, Delhi and M.P.

Das Gupta and co-workers (1939, 1940, 1941), Verma (1950, 1952) and Anon (1974-1985) confirm that no parasitic organism like bacteria or fungi are casually related to black tip of the mango. They do not find any virus or any deficiency related to the disease.

It is a popular belief that the disease is due to injurious fumes coming out of the brick kilns operating in the vicinity of mango orchards (Woodhouse, 1905 ; Das Gupta and Verma, 1939). This is further supported by the fact that the disease is not reported from the South India where modern Kilns of the type used but is reported in Northern India where such Kilns are not in use. Of the gases in coal fumes, sulphur dioxide, ethylene and carbon monoxide are especially toxic to the plants and of these sulphur dioxide causes the maximum damage (Rajan and Jha, 1940). The effect of SO_2 increases with increasing concentrations of gas and time of exposure (Sen *et al.*, 1943). The damage was found to be in inverse rates to the distance of orchard from Kiln. The maximum distance for the damage has been observed to be 700 yards but cases of damage upto a mile are also reported (Singh, 1960 ; Sen, 1943). Wind direction and velocity play an important role in increasing incidence of Black tip.

Symptoms of the disease in mango start appearing when the fruit is very small. The disease is noted in April-May when the fruits are only 1 cm long. No effect is seen on flowers. Fruit set is also normal. First symptom of the disease is the development of a small aetiolated area at the distal end of the fruit against the general green colour of the skin, which gradually spreads, turns nearly black and covers the tip completely. The tip is pulpy (although the rest of the fruit is compact, unripe and hard) and slightly sweet, but far different from the usual taste of the ripe fruit. The pericarp and mesocarp disintegrate exposing dark brown flesh beyond which the stone protrudes with a layer of collapsed tissue over it, the thickness of the layer depending upon the degree of distintegration the mesocarp tissue has undergone. Whole mango is never necrotic. The portion which is not directly affected is not of good taste and sometimes fail to ripen. Gummosis (exudation of drops of gums) is common in necrotic regions.

Fig. 17. Koeli (Black-tip) diseases of mango.

Control

Attempts have been made to cure the disease by spraying chemicals during the active fruit growth stage.

(*a*) *Prevention of black tip disease* can be made by :

(1) Restricting sites of new brick kilns at safe distances from orchards—about a mile east and west, and half mile on north and south direction (Pal and Chadha, 1980).

(2) The kiln should be stopped to operate between first week of March to third week of May (Srivastava, 1963 a, b).

(3) Brick kilns should be made practically smokeless by installation of certain pollution control fixtures to them (Blakke, 1913).

(4) To dilute the deleterious gases, the chimneys should be raised to atleast 40 to 50 ft. high or more.

(*b*) *Cure of Disease.*

(5) Use of borax and alkaline substances *viz.*, caustic soda and washing soda have been used to control the black tip. Since Borax undergoes alkaline hydrolysis in water

could also function as buffer against the harmful constituents of kiln fumes, which are generally acidic in nature. Perhaps, the presence of alkalinity on the fruit surface prevents entry of the deleterious fumes emanating from brick kilns in the form they are generated or they are changed or neutralized into a form which is less toxic.

(6) Spraying of borax (0.6%) thrice beginning from pea stage of fruit and subsequently at 15 days intervals (Yadav, 1997).

1.34. SPONGY TISSUE

It is a serious ripening problem in important commercial mango, Alphonso, the loss being as high as 30%. Sometimes mechanically injured fruits also develop spongy tissue like symptoms, where the exocarp may or may not be injured. X-ray photographs and X-ray images of fruits having spongy tissue show dark grey patches corresponding to internal cavities, in contrast to light grey areas of healthy flesh (Thomas *et al.* 1993) The affected parts of pulp becomes soft, spongy and with or without air pockets. It is suggested to harvest fruit little early (around a sp. Gravity of 1.0-1.02), keeping fruits under shade immediately after harvest or storing them at low temperature and encouraging growing of new hybrids free from spongy tissue (Yadav 1997). Chemical analysis of healthy and spongy tissue of Alphonso was done by Selvaraj *et al.* (2000). The breakdown tissue had low total carotene, sugars. Breakdown tissue had twice the amount of citric acid and half the quantity of malic acid. Ethylene forming enzyme activity was three-fold lower in breakdown tissues.

1.35. SOFT NOSE

Young (1958) from Florida (U.S.A.) has described this disease. Kent and Haden Mango are most susceptible to this disease. Physiological breakdown like 'spongy tissue' or 'soft centre' and tip pulp are reported by Krishnamurthy (1979), Subramanyam *et al.* (1971) and Verma (1950) ; but these are distinctly different disorders. The causal factors for the onset of this disorder are not known.

Five to twenty per cent fruits harvested from Haden and Kent cultivars are found affected but in severe cases, it may go upto 50 per cent.

In case of spongy tissue, symptoms are not apparent either at the time of picking or at ripe stage. The affected tissue is visible only when ripe fruit is cut into two halves. Such tissue is differentiated from the healthy one by its pale yellow colour, soft or spongy nature with or without air pockets and accompanied by off flavour (Subramanyam *et al.*, 1971). There is slight discolouration in the centre of the tissue surrounded by a soft halo. The 'Soft nose' is characterized by yellowing of the green skin in the area between the apex and the stigmal point. Lack of firmness is noticed in the diseased fruit, one can also feel with some experience. Upon cutting, the flesh on the ventral side towards the apex of some soft nose fruits merely appears to be over ripe, while that around the shoulders and on the dorsal side is unripe (Young, 1957).

Control

By developing growing resistant varieties.

1.36. INTERNAL NECROSIS

A browning of the flesh in mango fruit while still on the the trees has caused considerable loss to growers in Lucknow and other parts of U.P. The trouble, called Internal Necrosis, was first recognised by Ram (1976) and Anon (1974-85) in Deshehari mangoes which is different from black tip disease.

The first external symptom of the disease is the appearance of small isolated water soaked greyish areas on the fruit having the tip apparently healthy. In each of these greyish areas brown spots develop, which later on coalesce to form a dark brown necrotic area on the fruits, delimiting green parts. In severely infected fruits the complete disintegration of the pericarp and mesocarp of the diseased area occurs exposing the flesh which is dark brown in colour. Exudation of gum is seen along the diseased portion. Tissues of the lower part of the fruits are brown-black with corky and gummy pockets.

1.37. CLUSTERING

It is characterized by a cluster of fruitlets at the tip of the panicle giving an appearance of bunch tip called '*Jhumka*'. These fruitlets become dark green with a deeper curve in the sinus beak region than normal developing fruits. It occurs due to adverse climatic conditions during Feb.—March, when hermaphrodite flowers are more in number at the tip (Yadav 1997).

Control

Borax at the rate of 500 g per tree in soil or foliar spray (1%) or injection of (1%) borax significantly controlled the disease (Ram, 1976).

1.38. PHANEROGAMIC PARASITES AND EPITHYTES

Mango trees are commonly affected by parasites and epiphytes in neglected areas. They occur on the trunks and branches of trees. In North India, more than 60 to 90 per cent of mango trees are affected by these parasites (Singh, 1960). The yield of infected trees is reduced.

Among the parasites *Dendophthoe falcata* (L.F.) Ettings (*Loranthus*) a member of family Loranthaceae is most destructive and common pest of mango in India, which is distributed in wide areas. Besides this, five other plants of this family namely *Macrosolen cochinchinensis* van Tiegh., *Elytranthe capitellata* Engl., *Helicanthes elasticus* Dans., *Viscum articulatum* Burm. and *V. monocium* Roxb. parasitize the mango (Fischer, 1926). Of the above *Laranthus* and *Viscum* are common parasites on mango.

These plants produce haustoria which derive nutrition from the phloem tissue. Due to their luxurient growth the mango foliage is reduced resulting in reduced photosynthesis and thus less fruit setting. The fruits of Loranthaceae are baccate, with a viscous mesocarp and can be easily dispersed by the agencies like animals, birds (honey birds) squirrels and by explosion of their fruits (Singh, 1954). These seeds germinate and produce parasitic roots (haustoria). The penetrated host tissue sometimes swell to form tumours.

In areas of high rainfall, numerous epiphytes like flowering plants, ferns, mosses and lichens grow on the stems and branches of mango particularly on seedling types. *Vanda* sp., *Drimoglossum* sp. and several species of *Ficus* are reported on mango tree. Among the chief *Ficus* species are *F. parasitica* Koem, *F. lacor*

Buch-Ham. *F. rumphii* Blume, *F. benhalensis* L. and *F. religiosa* L. (Singh, 1960).

Control

(1) The affected branches should be cut and haustoria destroyed. The cut ends should be protected with Bordeaux paste or copper oxychloride spray to prevent the secondary pathogen infecting through the wounds.

(2) Injection of copper sulphate and Ferronoxone and 2-4D (2%) in the affected branches are quite effective (Kadambi, 1954).

(3) Spraying of diesel (30-40%) in soap water should be done as it was found effective by Singh (1960).

(4) An epiphytic plant should be removed at an early stage before it penetrates the roots in the ground and surround the host trunk.

2

Grapes

2.1. GENERAL ACCOUNT

Grapevine is the most important subtropical fruit in the world. 80 per cent of the crop is used for wine making, 10 per cent for making raisin and only 10 per cent is used as fresh fruit. In India it occupies only a small area. In Asia, the main centres are in Turkey, Israel, Syria, Iran and Afganistan. In India grapes used to come from Afghanistan, before independence. Grapes are cultivated in Japan and China too, but production here is limited to table grapes, raisins and wine for local consumption. India and Thailand are the chief producers of grapes. Other tropical countries with significant plantings are Taiwan, Colombia, Ethiopia, Ecuador, Phillipines and Venezuela (Winkler *et al.,* 1974).

Grape belongs to genera *Vitis* of family Vitaceae. It originated in the Caspian Sea area and came to India at a very early time. Commercial table grapes belong to the species *Vitis Vinifera* L. About 125 other species of *Vitis* are reported to be found growing wild in India. Out of these *V. barbata* found in Burma, Assam and Bengal, *V. rumeisperma* found in Assam and Sikkim and *V. parviflora* found in Kashmir, give edible fruits though of poor quality.

Commercial grape cultivation has assumed importance only during the past 20 years. Excellent vineyards now exist in Andhra Pradesh, Maharashtra, Madurai, and Bangalore in South. In Northern belt also in Delhi, Haryana, Punjab and Rajasthan grape cultivation is picking up. In our country 30,000 hectares are under grape cultivation with an annual production of 9 lakh tonnes (Arve 1992).

Grapes are mostly grown for table purpose in our country. As area under grapes increases substantially, it may cause glut in the market resulting in low profits to growers. It would, therefore, be necessary to lay emphasis on raisin grapes like Thompson seedless and Pusa seedless. Normally, the berry size in these varieties is somewhat small. This defect could be overcome by making raisins out of gibberellin-treated bunches as is being done in Russia.

Recently, some important raisin varieties like Kishmish Beli and Kishmish Charni (seedless) and Kata Kurghan and Nimrang (seeded) have been introduced from Russia and are under trial in different states.

The raisin grapes generally have 24 to 28 percent sugars, which include no sucrose. The sugar content of grapes grown in India varies from 13 to 22 percent. The sugars do not develop fully in humid weather of South and Central India. Attempts have been made to make raisins from grapes grown in Delhi by artificially treating them with sugar, but the product is leathery, sticky and of an indifferent quality. The vineyards in the Chini area (in Himachal Pradesh on Tibet border) are beginning to bear the fruits and it remains to be seen how successful it would be in supplying raisins.

Grapes can be grown on a wide range of climatic conditions from Himalayasto Madurai in South. However, the same variety does not do well under the different climatic conditions. For example, Anab-a-Shahi forms the basis of a booming industry in Hyderabad, where growth takes place all through the year, but it does poorly in North India, where there is a prolonged rest period during winter. The weather during the fruiting season affects the quality of fruit. Dry weather at the time of flowering is necessary for good fruit set. Rain or humid weather at the time of ripening causes splitting of berries and brings on an attack of fungal diseases. In North India, the problem is to get varieties which will ripen before the start of the monsoon. Summer temperature of 115°F causes sun-scales and grapes develop a thick skin. In Western India, there are two crops. One ripens in September-October after the monsoon rains, is of poorer quality and is often attacked by mildew. The other ripens in March and is sometimes damaged by occasional frost. The following grape varieties are recommended for commercial cultivation in different regions of the country (Randhawa, 1967).

Central India	— Anab-e-Shahi and Thompson seedless
South India	— Thompson seedless, Anab-e-Shahi and Bangalore blue.
Western India	— Anab-e-Shahi and Thompson seedless, Selection-7 Bhokri,Fakhri, Kali Sahebi, Pandri Sahebi, Gulabi and Bangalore Purple.
Northern India	— Khalibi, Perlette, Thompson seedless, Anab-e-Shahi (For dry belts only,) Kandhari, Beauty seedless and Black Prince.

Recently, Kishmish Beli has proved to be a promising seedless variety under Delhi conditions. This a Russian variety with high sugar content. Perlette is a famous Californian green seedless variety. It fetches a very high price as the fruits are brought very early in the market.

The most widely cultivated variety is Bhokri. It gives the highest yield, but the fruit quality is mediocre and the keeping quality is poor. The Berry is large, rounded and yellowish green. Fakhri is the second most important variety and has seeded berry. In Anab-e-Shahi the berries are very large, greenish yellow and firm have excellent keeping quality. Grape hybrids 'Arka Neelmani', Arka Shewta, Arka Majestic and Arka Chitra have been released in 1992-94 and Arka Krishna, Arka Soma land Arka Trishna were released in 1994 for juice and wine purpose by IIHR, Bangalore. Light soils are ideal for grapes. They can tolerate saline soils successfully. In grape cultivation, the most important and expensive operation is the training, pruning and staking of vines. In tropical climate pruning is done twice a year. The time of pruning determinate the fruiting. The common methods of training grapes all over the world are head system and kniffin or trellis system. However, in India the two commonest systems are the bower system and the single-stake system.

Grape is a tender watery berry and needs to be harvested and handled carefully. It is often shipped in bamboo baskets lined with soft grass. Now paper boxes containing 2 kg fruits are coming in vogue. These boxes are lined with newspaper cuttings or wood wool or paddy husk. Grapes must be harvested when fully ripe and when coloured varieties have developed their colours. Unlike mango and banana, they do not ripen after harvesting.

2.2. INSECT PESTS

2.2.1. Grapevine Flea beetle : *Scelodonta strigicollis M.*

(Family-Eumolpidae ; Order - Coleoptera).

The most important insect pest of grapes is the flea beetle, *S. strigicollis* found in Maharashtra and South India.

The adult beetle is small, reddish-brown with six black spots and metac. The adult beetles are very destructive as they nibble and eat away fresh leaf on vines after pruning. The grubs on the other hand are not very seroius as they feed on the cortical portion of roots. It has been reported to cause 11 percent damage (Nayar *et al.* 1981).

The female beetle lays the eggs in the soil or in the bark in groups or si.... The incubation period is four days. The grubs remain in the soil upto 15 cm and feed on roots and pupate in earthen cells. The adult beetles emerge in a wi..... Total life cycle is completed within two months.

Control

1. Spraying the vines, when fresh buds appear after pruning with DDT......per cent or parathion 0.025 percent is effective.
2. However, depending on severity of infection application must be repeated.
3. Lead arsenate can be used, or
4. A mixture of DDT and BHC can be sprayed.

2.2.2. Grape Phylloxera : *Phylloxera vitifoliae*

(Family - Phylloxeridae : Order-Hemiptera).

It is a well known destructive pest of grape. It is a native of middle and eastern United States.

The insect causes formation of galls on leaves and gall-like swellings on the roots. The European grapes or *Vinifera* varieties are highly susceptible to the pest. The insect got introduced into France in 1860 and about the same time it was also noticed in California. It is interesting to note that parthenogenetic females and sexually perfect females of *Phylloxera* sp. lay eggs.

2.2.3 The grapevine Stem girdler : *Sthenias grisator* Fb.

(Family-Cerambycidae : Order - Coleoptera)

It is mainly a pest of grapevine. But it can attack mango, jack fruit, *Morus indica*, *Cacao* and ornamentals like *Croton* and *Bougainvillea.*

The bark and wood are cut by powerful mandibles. In grapevine often the base of the vine is neatly grinded by the beetle which results in drying up of the entire vine. The beetle oviposits under the bark of girdled branch at night and the incubation period is about 8 days. The grub tunnels into the wilting branch. The whole life cycle completes within a year.

Control

Swabbing the base of main stem or branches with BHC 0.1 percent solution prevents infestation by the beetle.

2.2.4. The Castor semi-looper : *Achaea janata Linn.*

(Family - Noctuidae : Order - Lepidoptera)

It is one of the most important pests of *Castor* in India. It is a leaf defoliater and fruit piercing moth. Apart from *Castor* it feeds on grapes, pomegranate, *Zizyphus* sp. *Ficus benghalensis*, *Rosa indica* and *Bauhinia* sp. It occurs from August to January.

The eggs are round, blue green in colour and ridged. They are laid on the under surface of leaves singly and one to six eggs may be found on a leaf. A female is capable of laying about 450 eggs. The incubation period is 2-5 days. The larva is 60 to 65 mm with black head and the first pair of prolegs are non-functional and hence a semi-looper. Colour variations in larvae are also noticed. The larvae are gregarious and feed voraciously on leaves. It pupates in the soil or among fallen leaves. Rarely it pupates in leaf folds on the plant. Pupal stage lasts for 10-14 days in warm weather. The moth is pale reddish brown with hind wings black with a medial white band and three large white spots on outer margin.

Natural enemies

The Larvae are parasitised by a large number of insects like *Euplectrus leucostomus*, *Tetrastichus ophisuae* (Eulophidae), *Microplitis opphisuae* (Braconidae) etc.

Control

1. Hand picking and destruction of larvae will be useful.
2. Spraying of insecticides like parathion, endrin, toxaphene, carbaryl, dichlorvos and endosulfan may bring about appreciable control.

2.2.5. Udadya beetle

The grapvines are attacked by a large number of insect pests. *Udadya* beetle is one of them (Bindra, 1967). The beetles damage the foliage badly in the spring mostly at night.

Control

1. Removal of bark after prunning reduces the population of the *Udadya* beetle.
2. Spraying the foliage with 0.25 per cent DDT or BHC is recommended to control these beetles.

2.2.6. Ground beetle : *Gonocephalum depressum Th. andG. hoffmanseggi.*

(Family - Tenetrionidae : Order - Coleoptera)

These are large, elongate, somewhat flattened hard, reddish brown, often sculptured resembling superficially the Carabidae.

They are with vestigial wings. The elytra are frequently fused. Tarsi 5-5-4, heteromerous and claws are simple. These are nocturnal beetles and are pests of grapevine.

The tenebrionid *G. depressum* has been reported to be a vector of some of the larval helmiths of fowls, such as *Subuluro minetti*, the pinworm of fowl.

2.2.7. Leaf roller : *Sylepta lunalis* G.

(Family - Pyraustidae : Order - Lepidoptera)

Larva of *S. lunalis* roll the leaves of grapevine. They feed on green tissue in the early stage and eat up a large portion of leaf as they grow. Severe attack results in the presence of a large number of leaf rolls.

The moth is medium sized with yellowish wings having brown wavy markings. It lays pale white flat eggs singly on the under surface of leaves. Female may lay upto 200 eggs. The incubation period is 2-4 days. The larva becomes full grown in 15 to 20 days. It is 25 mm long and green coloured.

Control

Spray of endrin (0.02%) parathion (0.025%) carbaryl (0.1%) monocrotophos (0.04%) controls the pest effectively.

2.2.8. Berry scab thrip : *Scirtothrips dorsalis* Hood

(Family - Thripidae : Order - Thysanoptera)

Thrips are responsible for leaf curl diseases in chillies. They also occur on tomato, grapevine, mango and citrus.

They occur is such large numbers that they suck up the sap from the tender regions and cause the leaves to shrivel. In extreme cases the shoot hardly develops and the leaves fall off. It eats on fruit as well. Thus there are malformation of leaves, buds and fruits which may damage half the crop.

Fig. 18 .A—Healthy ; B—Diseased berries of grapes due to Periconia sarawatipurensis.

Control

Dusting or spraying BHC or spray application of dimethoate (0.03%) has been found effective.

2.2.9. Grapevine aleyrodid : *Aleurocanthus spiniferus* Q.

(Family - Aleyrodidae : Order : Hemiptera)

The aleyrodids are small insects with seven segmented antennae, two ocelli and a pair of reniform compound eyes. Aleyrodids are commonly called as white flies.

Sexual reproduction is the general rule but parthenogenesis occurs to some extent. Males develop from unfertilized eggs, whereas, from fertilized eggs both the sexes develop.

Control

Spray 0.05% fenitrothion, 0.05% parathion or 0.1% phosphamidon or dimethoate 0.03% to control the whitefly.

2.2.10. Berry Wasp : *Vespa orientalis Linn.*

(Family - Vespidae : Order - Hymenoptera)

V. orientalis belongs to vespinae (yellow jackets or hornet) are social wasps. The papery nests of these social wasps are constructed out of chewed foliage or wood. In the temperate regions the colony lasts for a season. This Queen over winters and commences formation of new colony in September. The nest is enclosed in a papery envelope. The nests are found in the open attached to branches of trees, etc. A few have their nests in the ground. The larvae are fed upon other insects, decayed or fresh meat.

2.2.11. Table : Some Other Insect Pests of Grapes

Sr.No.	Name	Family	Order
1.	*Odontotermes obesus R.* (White ant)	Termitidae	Isoptera
2.	*Coccus longulum D.*	Coccidae	Hemiptera
3.	*Pulvinaria maxima Gr.*	Coccidae	"
4.	*Helopeltis antonii S.*	Miridae	"
5.	*Retithrips syriacus Mayet*(Leaf thrips)	Thripidae	Thysanoptera
6.	*Florithrips traegardi Trybonis*(Flower thrips)	"	"
7.	*Thrips hawaiiensis Morgan*	"	"

VIRUS DISEASES

Grapevine suffers with a number of virus diseases. Of about 30 diseases, fan leaf, leaf roll and stem pitting are of economic importance.

2.3.1. Grapevine fan leaf

The disease occurs in form of trace fan leaf, yellow mosaic, vein banding and possibly enation. Detailed description is given by Vuittenez (1970). The disease is caused by a soil borne virus which is transmitted by the nematode *Xiphinema index* and *X. italiac*. Particles of the virus are isodiametric with angular outlines and about 30 nm in diameter.

Fan leaf can kill the vine under certain conditions of soil and climate. However, affected vines usually live long but become progressively unproductive. An average loss of yield of 50% had been reported for the most susceptible varieties. Berries become excessive short. The disease is restricted in nature to *Vitis* species probably due to narrow host range. Epidemiology and control of disease is discussed by Raski *et al.* (1983).

Grape fan leaf virus is detected by Serology which is done directly in the grapevine sap. Before serology this sap should be clarified and concentrated by centrifugation. The gel diffusion method of Ouchterlony (Shanmuganathan and Emmett, 1985/86) or Berck's (1967) latex agglutination technique is used. Shanmuganathan and Fletcher (1982) were able to detect this virus in dormant buds in winter, and in leaves, internodes and roots collected from glasshouse grown plants in spring by ELISA (enzyme-linked immunosorbent assay) technique. A number of workers have also used the ISEM (immunosorbent electron microscopy) technique for the rapid diagnosis of grape fan virus. Bovery *et al.* (1984) found both ELISA and ISEM equally effective in detecting this virus from June to October in leaf extracts of infected grapevine.

Control

Grapevine fan leaf virus is easily eliminated by heat therapy. Treatment of potted vines at 37-38°C for 30-40 days is sufficient to produce virus free shoot tips.

Barlass *et al.* (1982) developed a heat therapy technique in which apical fragments from virus infected vines are cultured *in vitro* in liquid medium at 27°C and then holding the culture at 35°C until new shoots regenerate. The generated shoots are excised and grown on a rooting medium. Vines produced in this way have been found to be free of virus during the first three years.

Since grape fan leaf virus spreads from diseased planting material, Virus-free planting material should be used.

Vector of grape fan leaf virus in not commonly found in tropics (Siddiqi 1974 ; Cohn, 1977), nematode control in vineyards is not a problem.

2.3.2. Grapevine Leaf roll

Vigour and yield of vines are reduced by reduction in size and number of clusters per vine. Fruit quality is reduced. The sugar content of the juice may be reduced upto 30%. The maturity of grapes on diseased vines may also be delayed by two to four weeks and berries of red variety may lack colour.

The disease can be readily identified in the field by the symptoms like black or red fruited varieties have purplish or reddish colour between veins in the leaves. This colouration first appears on the oldest leaves and subsequently on the younger leaves on the cane. The main veins usually remain green with a narrow band of green tissue on either side until leaf falls. In white varieties, the leaves turn yello prematurally but the main veins stay green. The berries become yellowish, giving the appearance of early maturity. Leaf roll is caused by a virus but despite a great deal of attention in the recent past, the etiology of the disease remains confused. However, evidence is mounting that the causal agent could be a member of the Closterovirus group. In Israel, Tanne *et al.* (1974 and 1977) isolated and characterized a filamentous virus (790 × 13nm) from leaf roll-infected grapevine. They assigned the virus to the poty-virus group.

Namba *et al.* (1979) from Japan and Tzeng and Goheer (1984) from USA concluded that a closterovirus is the probable causal agent of leafroll. There is no conclusive evidence of natural spread of leaf roll. The disease is spread by man through the use of infected propagation material of both root stocks and scion varieties.

Control

(1) By planting rootlings or graftings propagated from virus- free material.

(2) Leaf roll can be eliminated by heat therapy.

2.3.3. Grapevine stem pitting

The disease was first described as "legno riccio" in Italy, has been reported in several European countries Israel, California, Mexico and South Africa. It probably occurs world wide.

Affected plants show stunting, abnormal bark rugosity, and stem pitting. Bud burst is delayed on affected vines and there is an imbalance between scion androot stock growth.

In most cases stem pitting occurs when an infected scion variety is grafted on a sensitive root stock. Usually, Pitting symptoms develop on the stem and roots of the root stock upto the but union, but they can also develop on the scion variety. The disease is probably not the result of a scion root stock interaction because stem pitting has also been observed on ungrafted scion and root stock varieties.

Etiology of the disease is obscure. Conti *et al.* (1980) isolated a clostero-virus (800 nm) which is called as "stem- pitting-associated virus" by Shanmuganathan and Emmett (1985-86). There is no known vector for stem pitting. The disease can be detected by graft transmission to sensitive varieties such as Ohanez or *Vitis rupestris* du Lot. American workers (Hewitt, 1975 ; Teliz *et al.* 1980) believe that stem pitting is closely related to grape corky bark, another disease of unknown etiology, but this view is not supported by Italian studies of 'legno riccio'.

FUNGAL DISEASES

2.4. DOWNY MILDEW OF GRAPES

All the succulent tender parts of the vines are attacked, leading to a partial or total destruction of the foliage, dwarfing and killing of the shoots and rotting and cracking of the berries. The fungus spreads rapidly, with profuse downy growth on the undersurface and corresponding greenish patches on the upper surface, which turn chlorotic with age. The mildew growth may cover the entire leaf blade which turns brown and withers. The disease may also involve floral parts and the fruit. Infected flowers die and drop off. The fruit becomes greyish and the skin hardens and shrivels. Very often they are mummified while still attached to the bunch.

The disease is prevalent in all the grapevine areas of the world. It is serious throughout South India, especially in Andhra Pradesh and Karnataka states and is the chief limiting factor in grape production. Complete crop failures are usuallyencountered in South India whenever there is continuous rain during the early part of December. The disease spreads very fast when the rains are frequent with cloudy weather. Most of the commercial varieties except Bangalore Blue and Black Champa, are highly susceptible. Fourteen varieties out of 233 tested were found resistant at Bangalore (Sohi & Sridhar, 1971).

The pathogen responsible for grape downy mildew is *Plasmopara viticola* (Berkely and Curtis) Berlese and de Toni. The mycelium of the pathogen is intercellular with vascular haustoria. The pathogen produces large and branched sporangiophore with sporangia on small, rectangularly placed branches. The disease is carried from year to year in the form of oospores.

Control

1. Sanitary measures, such as cutting the primary source of infection by deep cultivation and ploughing, thus preventing the germination of oospore, are desirable practices.
2. Removing and burning the diseased leaves, shoots and berries that may contain hibernating oospores help in preventing epidemics.
3. The disease can be effectively controlled by giving three to five prophylactic sprays with 1% Bordeaux mixture.
4. Spraying with Fycol 8E (1%) gives good control.
5. Aureofungin and Chainin were effective. Ziram and Captan could only reduce the downy growth but the remaining fungicides were ineffective. Combination of Aureofungin with either Chainin or Streptomycin sulphate does not enhance the activity of two antibiotics (Kadkol and Gopalkrishnan, 1971).

2.5. POWDERY MILDEW OF GRAPEVINE

This is worldwide in distribution. The damage to vines is extensive in Europe and Western USA. There are frequent reports from Europe of the complete destruction of grape crop. In Afganistan, the loss in 1958 was estimated at 50-80% which is quivalent to Rs. 8.5 crore. In India this disease is common and destructive in Maharashtra, Gujarat, Tamil Nadu and Andhra Pradesh.

The pathogen responsible for this disease is *Uncinula necator* (Schw.) Burr. or *Oidium* sp.

It infects the plants at any stage of growth, attacking all the aerial parts, especially the young leaves and fruits. Leaves show white patches on both the surfaces. Diseased vines show restricted growth. Affected berries do not grow, their rinds become hard and ultimately crack. In case of early infection the fruits do not develop normally, become mis-shapen and crack.

Disease is more prominent before the onset of rains (May-June) in South India, whereas, it is more prominent during October-November under North Indian conditions. Disease development is retarded by sunshine and is favoured by warm sultry weather. Most of the cultivated varieties, especially Bangalore Blue, Bhokri and Anab-e-Shahi are highly susceptible. Apart from reduction in size and weight of affected berries, the disease reduces their market value. Upto 50% loss in yield is quite common in South India.

Control

1. Cultural practices including the trimming of developing shoots to reduce shading and allow free ventilation of vines are important for control (Ramakrishnan and Sundaram, 1955).

2. Pruning after the sheding of leaves, thinning out and cutting back of laterals, and removal and destruction of all diseased parts are important aspects of clean cultivation.
3. Dusting with sulphur is a universal practice for control in regions wheretemperatures are not high.
4. Bordeaux mixture and fixed coppers are also recommended (Anderson,1956).
5. Spraying the vines either with Sulfex (0.2%) or Bavistin (0.1%) gives effective control provided the first prophylactic spray is given at the proper time.
6. Control by Calixin (Tridemorph), Karathane (Dinocap) and Cosan with 4 sprays at 10 days intervals commencing from first appearance of symptoms (Gupta *et al.*, 1986).
7. Datar *et al.* (1979) studied the reaction of some grape varieties to powdery mildew. They found out of fifty seven varieties only Berlanderix riperia, Flame muscat, Large, White and Maloga to be resistant.

2.6. ANTHRACNOSE

The disease is caused by *Gloeosporium ampelophagum* Pass. which produces a perfect (sexual) stage viz., *Elsinoe ampelina* Shear = *Sphaceloma ampelinum* which belongs to family Elsinoaceae, Order-Myrinagiales of class Ascomycetes (Sohi, 1983). It is a serious disease prevalent almost throughout the country and is a limiting factor in Andhra Pradesh, Karnataka and Tamil Nadu. It is serious on new sprouts during the rainy weather. Under North Indian conditions, the disease appears only during the monsoon season. The berries escape infection, if the crop matures before the onset of rains. Under South Indian conditions, all the aerial young parts of the plant are attacked, especially the leaves.

Leaves produce dark brown spots during July-August and November-December. Depressed cankers are produced on the stems and twigs. The growth of the young infected twigs is arrested, which remain short and eventually die up. On the berries, circular brown sunken spots with dark brown margins develop leading to their shrivelling and drying. Serious crop losses are experienced in South India on variety Anab-e-Shahi. Varieties Pusa Seedless, Chandigarh, Delight, Perlette, Romulus, Motia, Black Muscat, Fakdi, Hubshi, Hur, Thompson Seedless behaved as highly susceptible at Saharanpur (U.P.) whereas, Himrod, Schuyler White, Beauty Seedless, Niabel, Muscat, St. Valior, Large White, Golden Queen, Bangalore Blue, Isabella and Golden Muscat were resistant (Prasad and Nirvan, 1965). Some new hybrids (A. Shyam) have been released from Bangalore using Bangalore Blue and Black Champa as resistant donar parents.

Possible role of sugar in anthracnose disease resistance mechanism in two grapevine varieties Anab-e-Shahi and Bangalore Blue was studied by Mohanraj *et al.* (1972), They reported that grapevine anthracnose is a high sugar disease. Unlike other Anthracnose disease which are low sugar diseases.

Control

1. Removal of diseased twigs.
2. Destruction of the pruned materials.
3. Spraying of the vines immediately after pruning with Bordeaux mixture (5:5:50) and fortnightly sprays with Bavistin (0.1%) of Difolatan (0.36%) or Bordeaux mixture (3:3:50) during the rainy season have consistently given good control under Bangalore conditions (Sohi and Sridhar, 1972).
4. Since the build up of inoculum is very fast during the rainy weather, repeated sprays at shorter intervals (3-5 days) are essential.

2.7. RUST OF GRAPES

It is caused by *Phakospora ampelophagu.* It usually attacks the Labrusca group of grapes (Sohi, 1983). Variety Bangalore Blue is very severely attacked.

The symptoms are numerous, orange coloured sori mostly on the lower surface of the leaves. Severely infected older leaves show upward curling, distortion and drying up. The fungus produces only the uredo stage. The incubation period of the disease on detached grapevine leaves was 5-6 days at 16-30°C and 15-20 days at 12°C in laboratory. Uredospores germinated at 100 Ft. candles or less but not at 200. Darkness favoured the germination. Optimum temperature for germination was 24°C and RH 75.3% or above was favourable. Infection pegs penetrated through the stomatal aperture after 12 hr. of inoculation and the leaf tissue was colonized within 5 days.

Uredia are initiated 6-7 days after inoculation (Leu and Wu, 1983).

Control

Spraying with Sulfex (0.2%) or Sulphur dusting during morning hours is helpful in controlling it.

2.8. PREMATURE DECAY OF GRAPEVINE

Streum hirsutum (Willd.) Fr. occurs sporadically on grapevine in Hungary, but in 1982, it caused considerable infection (11.6%) on cv. Red Traminer (Lehoezky and Mako, 1983).

Symptoms first appear on leaves, which fall prematurely. Clusters wilt and infected vines decay in next summer. Sometimes a sudden wilt occurs. Brown discolouration and white spongy rot affects the xylem. Sporophores develop rarely on the woody parts of infected plants. Infection is favoured by rainy wealther and poor aeration.

Control

Diseased parts should be early diagnosed and they be removed before the mycelium reaches the graft union.

2.9. WILT OF GRAPEVINE

Fusarium oxysporum Schlecht. and *Cylindrocarpon destructans* were found associated with the affected tissue of the bark and wood from Sicily (Grasso, 1984).

Symptoms of the disease are leaf yellowing and wilt and cane blight, accompanied by necrosis of cortical tissues and browning and tylosing of the xylem, starting from the collar. Inoculation by root immersion induced browning of the xylem and reduced plant growth, especially at the expense of the root system and the fungi were reisolated one year later.

2.10. BITTER ROT OF GRAPES

Prakash *et al.* (1974) reported bitter rot of grapes caused by *Greeneria fuliginea*Scribner and Viala. This notorious fungus is referred as *Melanconium fuligineum* (Ser. and Viala) Cav. from U.S.A. (Luttrell 1953, Luttrell and Murphy, 1953) and was responsible for bitter rot or American rot of grapes. The fungus was recovered from Anab-e-Shahi variety of grapes in September 1971 from Dholi, Muzzaffarpur, Bihar. Reddy and Riddy (1893) described symptoms and pathogenicity trial results.

Fungus attacks unripe berries just near the peduncle. In some cases entire branches are infected. The berries often drop down when the whole peduncle is infected. The affected berries are light brown to dark fulginious, full of dense acervuli but they do not shrivel. On young barries, superficial circular brown spots develop. In severe cases of infection, the discoloured areas coalesced and spread rapidly all over the fruit. Black acervuli of the fungus developed in the discoloured areas. Girdling of the canes was observed to be quite common. Bark cracked in longitudinal fashion at the girdled region. Such exposed wood appeared brownish in colour. Disease may show symptoms on leaves as well. The withering of the leaves starts from down upwards. Such withered leaves never turned yellow, but finally became cinnamon brown in colour and remained attached to the canes until they were disturbed mechanically.

2.11. PERICONIA ROT

A soft rot of grapes caused by *Periconia saraswatipurensis* Bilgrami was reported by Rai (1983) from Allahabad (U.P.).

The diseased grape berries showed water-soaked, slightly depressed spots which enlarged and become dark brown with the march of the fungus. White to pinkish mycelia of the pathogen along with fruiting bodies appeared on the centre of the diseased area and covered the fruit completely within 8-9 days.

The rotted fruits emitted fermented odour and the affected tissues became pulpy.

P. saraswatipurensis could produce infection on the fruits of *Carica papaya*, *Citrus reticulata*, *Physalis peruviana*, *Prunus domestica* and *Prunus persica*, while the fruits of *Achras sapota*, *Aegle marmelos*, *Citrus sinensis*, *Emblica officinalis*, *Litchi chinensis*, *Malus pumila, Mangifera indica, Musa paradisiaca, Psidium guajava* and *Zizyphus mauritiana* were resistant to the fungus.

Percentage spore germination was best on host extract (Berry extract) 94% followed by Basal medium (Asthana and Hawker's modified medium A) 83% and in 1% glucose, 78% germination. Of the various temperatures used 25°C gave 77% and 30°C produced germ tubes in 70% spores. Thermal death point was 50°C. At 5.5. potentials of Hydrogen 79% germination was observed which was 72% at Ph 6.0.

For growth and sporulation best pH was 5.5 and temp. 25°C. Growth was good on D(+) Xylose, Lactose, Dextrin and Fumaric acid. Urea and Potassium nitrate were suitable nitrogen sources.

In healthy grapes glucose and fructose were detected. During pathogenesis caused by *P. saraswatipurensis* these sugars gradually depleted and their complete disappearance was recorded on 12th day. The percentage loss of ascorbic acid in grape fruits infected with the pathogen was 81.0% after 12th day while the corresponding loss in healthy fruits was 72.5%.

Control

1. Out of different chemical treatments of fruit, percentage rot of grapes due to infection caused by *P.saraswatipurensis* was least at 50 ppm of Bavistin followed by Benlate and Difolatan at 100 ppm followed by Benlate and Difolatan at 100 ppm, TBZ at 150 ppm and Vitavax at 750 ppm (Rai, 1983).
2. Among antibiotics Aureofungin was effective.
3. Various preventive measures were suggested by Rai (1983) such as careful handling and exclusion of infected leaves in the packing, use of clean containers, quick transport and sales.
4. Storage of fruits at low temperature (5 to 10°C) is suggested.

2.12. PHOMOPSIS ROT OR DEAD ARM DISEASE

The dead-arm fungus *Phomopsis viticola* Sacc. was observed by Saccardo as early as 1880 and since then it has been reported as a severe pathogen of grapevine (Bolay *et al.*, 1968 ; Costa Maria and Camara, 1953 ; Cucuzza and Sall, 1982 ; Reddick, 1909 ; Shea, 1961) from various grape growing areas of the world. The disease infecting shoots of the vine is commonly known as dead arm disease of grapevine besides some other less common names viz. necrosis of grapevine, root neck blight, excoriosis of grapevine and black spot disease of grapevine. This disease was reported from India by Lal and Arya (1982) from Allahabad, Kanpur and Fatehpur districts of U.P.

The causal fungus over winters in the outer bark of bud spurs infected in the previous growing season. Pycnidia develop in the bark tissue. High summer temperature suppresses activity of fungus. Symptoms of disease are most prominent on the basal leaves and lower portion of the shoot. Leaf symptoms consist of pin point, reddish brown spots surrounded by a zone of chlorotic tissue creating a halo effect. If lesions are numerous the leaf becomes distorted and tattered ; occasionally the blade falls off at the petiolar junction.

Fig. 19. Phomosis rot of grapes. (1-A)—Healthy, (1- B)—Diseased ; 2—Pycnidia ; 3 & 4—Kali sahebi and Bhokri varieties showing disease symptoms.

On shoots, symptoms appear as small brownish black spots that may expand to oval-shaped lesions. If numerous lesions occur at the base, they often coalesce and blacken that portion of the shoot. The vine is damaged primarily by death of severely infected shoots.

The grape berries initially developed small brown coloured spots which later on became round, with distinct dark brown margins. Infected tissue became pulpy,and white mycelial growth appeared after 4-5 days of incubation at 25-28°C. The berries completely rotted within 10-12 days and emitted a fermented odour. Black, pinhead like fruiting bodies (Pycnidia, characteristic of the disease) developed over entire surface of the fruit after 12 days.

Out of the three varieties tested for disease resistance Bhokri (Nasik green) a south Indian grape variety was comparatively more resistant than the Thompson Seedless and Kali Sahebi varieties. Percentage rot was 80.2% in Bhokri after 15 days as compared to 100 per cent in other two varieties. Utilization of seven alcohols was studied using chromatographic technique (Ranjan *et al.,* 1955). Sorbitol produced maximum growth while no growth was observed on methanol, ethanol and ethylene glycol (Arya and Lal, 1985). *P.viticola* produced best growth and excellent sporulation in proline (out of 14 aminoacids). It more efficiently utilized glycine, aspartic acid and tyrosine within 13 days. In a study, a mixture of 4 amino acids was taken. It was interesting to note that *P.viticola* failed to consume aspartic acid from mixture 1 and glycine from mixture 3 both of which were consumed in 13 days when suplied individually. In mixture 2, preferential utilization of serine and lysine was observed by this pathogen (Arya, 1987).

Biochemical changes in grapes were studied by Arya and Lal (1986). Due to infection the quantity of free amino acids decreased, the asparagine was completely utilized and valine was detected only in traces. On the contrary the amount of total bound amino acids increased. Similarly diseased berries showed a pronounced increase in organic acid contents after 10 days, but the amount decreased ultimately. A constant decrease in sucrose, glucose and fructose was observed. An unidentified oligosaccharide was observed in healthy as well as diseased fruits but it was depleted after 5 days. Percentage loss of Vit. C was observed in three grape varieties viz., Bhokri, Thompson seedless and Kali Sahebi. It was 83.2, 88.5 and 84% after 10 days of infection while corresponding loss in healthy berries was 36, 42.3 and 73.3% in three varieties respectively (Arya and Lal, 1986).

Control

1. The dormant treatment with sodium arsenite and dinoseb significantly reduced the number of active pycnidia. The foliar treatment with Captan was not effective (Cucuzza and Sall, 1982). Bolay and Pezet (1987) recommended avoidance of excess copper in controlling various grape diseases including *P.viticola.*
2. Satisfactory control of fruit rot was obtained by Bavistin, Difolatan, Calyxin (1000 ppm each), Dithane M-45 (1250 ppm). Jaillowx *et al.* (1987) recommended use of Fosety1-A1 + Folpet for control of this disease. Excoriosis causes serious damage also in temperate zones with a mild rainy spring. Certain preventive measures were

suggested by Carles (1984) : (a) Dipping the bud wood for 15 h at 20°C in a hydroxiquinoline sulphate sol. at 5g/lit. (b) Sodium arsenite 625g/lit. applied at 500 *l*/ha may be used at 2-3 weeks after pruning and 2 weeks before bud burst.

3. Of the various extracts, *Eucalyptus* and *Aegle marmelos* Corr. (50%) was found effective (Arya, 1988 a). 75% leaf extract of Kuchila (*Strychnos nux-vomica* L.) controlled 55% rot of grapes. Control by *Eucalyptus occidentalis* Skeel and Neem (*Azadirachta indica* A. Juess.) was also effective (Arya, 1988b).
4. Amongst antibiotics Idifulvin (1500 ppm) and fixed oils, mustard oil were most effective.
5. As a preventive measure, injury to the ripe berries should be avoided and regular sorting should be practiced. Infected berries which serve as potential source of secondary inocula during transportation and storage should be immediately destroyed. The cross inoculation studies revealed that the fungus can grow luxuriently on a number of other hosts, direct contact of infected berries with such fruits and vegetables should be avoided as far as possible.

2.13. PESTALOTIA ROT

Mishra *et al.* (1974) reported a severe and widespread rotting of grape berries in the vineyard of College of Agriculture, Dholi (Muzaffarpur), Bihar during August, 1970 on Anab-e-Shahi and Muskat varieties.

The infection starts just near the peduncle or the tip of the fruit, subsequently, it covers the upper portions of the fruit within two days. First the lesion appears as water soaked and later turns Sienn's colour with numerous acervuli of the fungus. Infection starts when fruits are about to ripe. Lesions are irregular, depressed in size and shape associated with raised acervuli in severe cases. The skin of the fruits become brownish black and leathery, lighter in weight and completely unmarketable. The disease also appears on twigs and leaves of the vineyard. Macroscopic and microscopic examination of diseased material revealed the presence of *Pestalotia menezesiana* Bres. & Torr.

Raw, semi-ripe and fully ripe berries (15 each) were inoculated to observe the percentage rot. Semi-ripened and picked berries gave more infection (93.3 per cent) follwed by fully ripened (55.3 per cent) and raw (20.0 per cent). Unpicked fruits had less percentage of infection in each case.

The morphological characters of the fungus agree with *P.menezesiana* reported by Mundkur and Thirumalachar (1947). Sohi and Prakash (1972) also reported *P.univola* on grape berries from Bangalore.

2.14. GREY MOULD

Grey mould or Botryis rot is caused by *Botrytis cinerea* Pers. (Anamorph : *Sclerotinia fuckeliana* (de Bary) Fuckel. Under favourable conditions, this fungus attacks senescent or fleshy plant tissues with high sugar contents such as floral parts or ripening fruit. With grapes, grey mould is frequently responsible for the rot of bunches either at or just prior to harvest or, after harvest while fruit is in transit or cold storage.

Cultivars of *V.vinifera* and other *Vitis* species vary in susceptibility to grey mould. Generally white grape cultivars that develop tight clusters are more prone to pre-harvest attack (Sall *et al.,* 1981).

Berries affected by grey mould, at first, develop a soft brown rot. As the disease progresses, grey spore masses are produced at cracks in berry skin and later over the entire surface of berries. Clusters of adjacent berries are subsequently infected, rotted and matted together.

Fig. 20. Grey mould or Botrytis rot of grapes.

B.cinerea survives from year to year on grapevines as sclerotia either on or within canes (Nair, 1985 b) or rotten berries and stem clusters. Frequently, sclerotia persist in grape mummies left on vines from previous harvests (Sall *et al.,* 1981). In damp conditions, conidia are produced on the surface of sclerotia and spread by wind to sites of potential infection. In localities where cool wet conditions persist for long periods, young leaves, shoot, stems and developing flower clusters are occasionally attacked and rotted.

Control

Grey mould control involves the application of well timed fungicide treatments integrated with vineyard sanitation and vine canopy management practices. Sources of *Botrytis* inoculum within vines can be reduced by removing grape mummies from vines at prunning and burying them. As moisture and temperature on surfaces of flowers and berries influence grey mould development, changes to vine that favour drying of vine foliage also contribute to disease control (Sall *et al.,* 1981 ; Savage and Sall, 1984).

In chemical treatment programs, fungicides are usually applied during or just after flowering to prevent cluster contamination and to protect developing berries from infection and at cluster closure and before harvest to minimise the spread of disease onto ripening berries. Recent studies have also indicated that pre-flowering treatments reduce levels of latent infection in flowers and developing berry clusters (Nair, 1985a). Effective fungicides include captan, benomyl (Winkler *et al.,* 1974 ; Sall *et al.,* 1981), iprodione, vinchloizolin and procymidone (Emmett *et al.,* 1982). 200 ppm SO_2 controlled the development of disease in inoculated berries (Marois *et al.,* 1986). Application of dicarboximides reduced the infection of flowers and berries (Nair and Nadtotchei, 1987). Procymidone, vinchlozolin and iprodione at 1g/*l* were found effective by Nair (1987). *B.cinerea* isolated from grapes was treated with SO_2 diffusing from polyethylene bags (of various sizes and thickness) containing sodium metabisulphite. Best control was achieved in 9 × 8 Cm_2 polyethylene bag 80 or 100 μm thick at 1°C with 90% RH (Boubekri, 1987). The efficacy of procymidone at 50, 75, 100 a.i./litre was compared with iprodione (100g/100 litre) and tiofanato metilico (thiophanate- methyl, 49g a.i./ litre) for the control of *Botrytis cinerea* (Menten *et al.* 1997).

Selection of antagonistic strains of *Trichoderma* and determination of conditions of application in the field was done by Dubos (1986) in France, against grey mould of gragpevine. Vinclozolin + Thiram were found effective by Chichignoud and Lartaud (1986).

2.15. CERCOSPORA LEAF SPOT

Cercospora leaf-spot, also called Isariopsis blight, is caused by *Mycosphaerella personata* Higgins (anamorph : *Cercospora viticola* (Ces.) Winter and Demetrio ; Syn.*C.vitis* (Lev.) Sacc. or *Isariopsis clavispora* (Berk. & Curt.) Sacc.

Cercospora leaf spot had been recorded on cultivars of *V.vinifera* and other *Vitis* species in countries such as U.S.A. (USDA, 1960), India (Singh *et al.,* 1976) and NewZealand (Harvey and Wenham, 1972).

Munjal and Sethi (1966) found two species of *Cercospora* on leaves of grapes. *C.viticola* is the most prevalent species causing leaf spot. Leaf spots are circular, rarely irregular, 5-10 mm in diameter, dark brown with an ashy-grey centre. *C.riachueli* Speg. is another species, here leaf spots are irregular 2-8 mm in diameter, dirty brown to tan coloured. Studies on the nutritional requirements of *C.viticola* were done by Sethi and Munjal (1963).

Control

Fungicides used for the control of Cercospora leaf spot include benomyl, mancozeb, captafol and chlorathalonil (Hopkins, 1974 ; Bose *et al.,* 1975).

2.16. BLACK ROT

The disease is caused by *Guignardia bidwelli* (Ell.) Viala and Ravaz. (anamorph : *Phyllosticta ampelicida* (Enleman van der Aa.) which is indigenous to Eastern North

America. Black rot is a destructive disease on cultivars of *Vitis vinifera* and some other *Vitis* species (McGrew and Still, 1972 ; Winkler *et al.,* 1974 ; Sivanesan and Holliday, 1981).

G.bidwelli attacks all green parts of the vine, particularly berry clusters. On leaves, red-brown irregularly shaped spots develop in which black fruiting structures of the fungus are produced in concentric rings. Small dark elliptical cankers are produced on shoots, tendrils, cluster stems and petioles. First symptoms on berries are light brown soft circular spots. As these enlarge, entire berries become discoloured. Affected berries change to hard black wrinkled mummies covered with pycnidia of the fungus. The mummies remain attached to cluster stems (McGrew and Still, 1972). Biology of the fungus was described by Barbe (1984).

Control

Control is based on removal of sources of potential inoculum and the application of protectant fungicides.

Destruction of infected vine material (prunings and mummified fruits) by burning or by deep burial helps to reduce inoculum levels.

Pre-bloom and post-bloom treatments are done with fungicides like Bordeaux mixture, Folpet and Maneb (Sivanesan and Holliday, 1981), benomyl, captafol and captan (Hopkins, 1974), Fenarimol and Zinc-maneb or mancozeb (Spotts, 1979), Ferbam (Mc Grew and Still, 1972) Captafal (180 g/h 1) Captan (175g/h1), dichlofluanid (250 g/h1) Folpet (175g/h1) maneb and propineb (280g/h1 each) and Zineb (250 g/h1) (Barbe, 1984).

2.17. WHITE ROT

Grapevine white rot, caused by *Coniella diplodiella* (Speg.) Petrak & Sydow (Syn. *Coniothyrium diplodiella* (Speg.) Sacc.) occurs in many parts of the world such as Europe, North and South America, Asia, Africa and Australia including some subtropical regions. It is sometimes referred to as 'hail rot' because severe outbreaks often occur after hail storms (Anon, 1966).

Although all green parts can be attacked, clusters are usually damaged more than shoots, stems and leaves. When cluster stems are affected, brownish depressed cankers develop and many enlarge until the stems are girdled. The entire cluster beyond the girdle eventually withers and dies. Infected berries become juicy before turning a white grey colour and drying out. Brown-black pycnidia of the fungus are eventually produced under the skin of withered berries.

C.diplodiella can survive for 2-3 years within pycnidia on vine debris such as withered berries both in soil and on the vine (Chrelashvili, 1972). During rain or hail storms, spores are splashed onto vine foliage and grape clusters. The fungus can also be spread by implements used for cutting bunches. Infection takes place through wounds resulting from damage by hail, sun scroch, mechanical wounding or attack by downy mildew (Anon, 1966) or through guttation pores on leaves (Chen *et al.,* 1980). During period of high relative humidity, exosmosis of nutrient from wounded tissues stimulates spore germination and infection. Grape clusters near to or on the ground are more prone to attack (Chen, *et al.*, 1979, 1980).

Control

By use of fungicides such as Captan and Phthalan can reduce the infection (Chrelashvili, 1972).

2.18. GLOMERELLA ROT

The disease is referred to as ripe rot (Shanmuganathan and Emmett 1985/86) and is caused by *Glomerella cingulata* (Stonem) Spaulding & Schrenk (anamorph: *Colletotrichum gloeosporioids* (Penz.) Sacc. Although this fungus has a world wide distribution, it occurs more abundantly in the tropics and sub-tropics than in temperate regions (Mordue, 1971).

Symptoms include small sunken spots which are pale red to brown in colour on white fruited cultivars. As affected areas on fruit gradually increase in size, concentric rings of darker coloured acervuli develop. These eventually rupture the skins of infected berries.

Development of *G. cingulata* is favoured by conditions of high moisture and temperature. Growth is optimum at 29°C while spore germination, infection and ascospore production and discharge require relative humidities approaching 100%.

G.cingulata persists from year to year in berry mummies, pedicels on fruit spurs. Water-borne conidia are then dispersed onto developing berries during rains. Infection of fruit occurs at all stages of development and remains latent until ripening (Daykin and Milholland, 1984).

Control

Fungicides such as Captafol, Captan, Maneb and Benomyl control ripe rot when applied at 14 day intervals from bloom until harvest (Daykin and Milholland, 1984).

2.19. PENICILLIUM ROT

Tandon *et al.* (1975) while reporting some diseases of fruits and vegetables from Nainital, Bhowali and their neighbouring areas in June 1974 found *Penicillium citrinum* Thom. causing severe deterioration of grape berries. Sharma and Dharamvir (1985) found *P.canescens* Sopp., causing fruit decay in Delhi market *P.chrysogenum* is also found on grapes (Singh and Kainsa, 1983).

The infection started from skin cracks or an injury to the fruit. The healthy fruits in the bunches soon became infected when they came in contact with the diseased ones. A dirty olive green growth of the organism appeared on the infected region. About 20% decay due to *P.citrinum* occurred in the individual fruit and the intensity of the disease ranged between 11 to 14 per cent.

Control

1. Tandon *et al.* (1977) tried four fungicides, calixin, cupramar, ziride and bavistin. Only bavistin could control the rot upto 95, 90 and 81 per cent when used at 2000, 1000 and 500 ppm respectively.
2. Earlier Tandon *et al.* (1975) had recommended captan for the control of the decay of grape berries caused by *A.niger* and *Penicillium* sp. They reported 70 and 80% control at 2000 ppm.

2.20. ALTERNARIA LEAF SPOT

Suhag and Kaushik (1982) reported *Alternaria vitis* (Cav.) Sacc., on the leaves of grapes. Goyal *et al.* (1971) found *A. alternata* (Fr.) Keissler associated with the leaves from Rajasthan.

A.vitis produced large irregular yellowish disease spots having concentric rings, such patches enlarged and coalesced rapidly covering a large area of leaf lamina. However, leaf veinal system remained intact.

2.21. SPHACELOMA LEAF SPOT

Sphaceloma ampelinum de Bary produced dark brown lesions which spręad characteristically along the leaf veinal system followed by complete distortion of eaf lamina (Suhag and Kaushik, 1982). Rangaswamy and associates (1970) reported the pathogen from Dharwar in South India.

2.22. ASPERGILLUS ROT

Microbial population of freshly harvested Perlette and Delight grape berries was studied by Singh and Kainsa (1983). They found association of *A.niger* Van Tiegh. *A.rugulosus* Thom and Raper, *A.terreus* Thom on both the cultivars during growing season. Jarvis and Traquair (1984) reported bunch rot of grapes caused by *A.aculeatus* Lizuka from South West Ontario. *A.carbonarius* was reported by Hegde and Kulkarni (1985).

A. niger is a severe pathogen, it causes typical stalk rot symptoms. The pathogen gains entry through the minute cracks near the stalk and causes water soaked lesions which ultimately cover the whole fruit. The fungus produces black conidial heads after 8 days of infection.

A.aculeatus also enters through the fractures caused by partial detachment at the pedicel when the grapes were in tight bunches, and through splits and insect punctures. This disease is associated with high temperature. Cleistothecia were formed in culture but were not seen in the field or in inoculated grapes.

Control

1. *A.niger* was best controlled by diphenyl used as a protectant and eradicant. Stable bleaching powder was second of the 4 chemicals tested while sodium metabisulphate was good protectant but not eradicant (Singh *et al.*, 1985).

2. No infection upto 4 days except on 2-4D. Disease was not observed upto 8 days of incubation in Carbendazim, Difolatan and VFI paper wraps. Wrap treatment showed least decrease followed by Carbendazim. The disease incidence was at par in Difolatan than in Thiabendazole.
3. Tandon *et al.* (1977) found Calixin ineffective against *A.niger*. Bavistin, Cupramar and Ziride at 2000 ppm checked the rot upto 90, 60 and 50 per cent. Bavistin controlled 85 and 72 per cent disease at 1000 and 500 ppm, respectively. Other two fungicides were less effective at lower concentrations.

2.23 TABLE : SOME MORE FUNGI ISOLATED FROM GRAPEVINE

Parts used for Isolation	Organism	Reference
Root	*Verticillium dahliae*	Moller, 1981 a.
Root	*Phytophthora cinnamomi*	Moller, 1981 b.
Root	*Armillaria Sp.*	Moller, 1981 c.
Root	*Pythium sp.*	Winkler, *et al.*, 1974.
Stem	*Eutypalata (syn.E.armeniacae,E utypa die back)*	Moller & Kasimatis, 1981.
Root & Stem	*Thielaviopsis basicola*	Canter-Visscher and Over de Linden, 1972.
Leaf	*Drechslera rostrata*	
Leaf	*Botryodiplodia palmarum*	Hedge and Kulkarni, 1985.
Leaf	*Hendersonula torulodies*	
Leaf	*Septoria ampelina*	Boubals and Mur, 1983.
Leaf & Stem	*Diplodia natalensis*	Strobel and Hewitt, 1964.
Fruit	*Rhizopus nigricans*	Winkler *et al., 1974.*
Fruit	*R. arrhizus*	Sall *et al., 1981.*
Fruit	*R. oryzae*	Singh and Kainsa,, 1983.
Fruit	*Cladosporium herbarum*	Winkler *et al., 1974.*
Fruit	*C.cladosporioides*	Nair,, 1985 a.
Fruit	*Stemphylium sp.*	Winkler *et al., 1974.*
Fruit	(Sour rot) *Saccharomyces sp.*	Singh and Kainsa,, 1983.

2.24. BACTERIAL DISEASES

2.24.1. Crown gall

Crown gall caused by *Agrobacterium tumefaciens* is most widespread bacterial disease of grapevine. The disease development is favoured by wet conditions and high temperature. Disease performance varies according to cultivar and age of vines (Winkler *et al.,* 1974).

2.24.2. Bacterial blight

Disease is caused by *Xanthomonas ampelina*, which occurs in Europe and South Africa. Foilage and developing clusters are attacked and the frame of vines becomes infected. In severe cases vines may die.

2.24.3. Pierce's disease

Pierce's disease is a destructive disease in some parts of North America. For sometime the cause of this disease was thought to be a rickettsia like organism (Goheen *et al.,* 1973). But it is now considered to be caused by a bacterium (APR, 1985), cultivars of *Vitis vinifera* are very susceptible to Pierce's disease. Some American *Vitis* species are resistant (Winkler *et al.,* 1974).

2.24.4. Other diseases

Other bacterial diseases affecting grapevines include *Xanthomonas vitis-vinifera* and *Pseudomonas viticola* both of which cause leaf spots in India (Nayudu 1972 ; Hegde and Kulkarni 1985). *Pseudomonas viridiflava* is reported from New Zealand (Wilkie *et al.,* 1973). Various bacteria are associated with pungent sour rots. For example, *Acetobacter* sp. (Sall *et al.,* 1981).

3

Guava

3.1. GENERAL ACCOUNT

Guava (*Psidium guajava* L.) is the fourth most important fruit in India in area and production (Ranjit Singh, 1969). It originated in tropical America and seems to have been growing from Mexico to Peru when European explorers first visited the region. Though the exact time of introduction of Guava in India is not known ; but it must have been at a very early date, as it is mentioned by Briton (cf. Hayes, 1960), who was in India early in the 17th century.

Botanically, guava belongs to the large family Myrtaceae. This family includes the Eucalyptus tree, Cloves, Nutmeg, Allspice, and also the Jamun, Jambolan or Java plum (*Syzygium cumini* (L.) Skeels). Guava grows in a wide area in India, but guava grown in Allahabad are famous all over the world for the fine quality of the fruit. It is a sub-tropical and tropical fruit which requires a distinct winter for developing good fruit. It does not thrive, if the annual rainfall is more than 250 cm. The importance of guava is due to the fact that it is a hardy fruit which can be grown in poor alkaline or poorly-drained soils without any manuring or irrigation. It can grow in soils with the pH ranging from 4.5 to 7.5.

Apart from cultivated guava, some other allied species with edible fruits are also found in India. But they are no commercially important. The Strawberry guava or Cattley guava (*P. cattleianum*) has purplish red fruits about 2.5 cm in diameter. It has a yellow fruited form (var. *lucidum*), which grows in South India. The Guinea guava or Brazilian guava (*P. guinese*) has a small fruit of poor quality and rounded branches. The young branchlets in the ordinary guava are four sided. The feijoa or pineapple guava (*Feijoa sellowiana*) is grown on the hills of South India. It is also reported to be growing in Bengal.

There is a great variety in the types of fruit. Apple guavas have a brilliant red blush. In Kushroo baug, Allahabad presence of Mn in soil is said to be responsible for red colour.

Some types are pear-shaped and some even look like a bitter gourd or 'Karela'. The most famous variety is the Allahabad Safeda with a round smooth fruit and white flesh of excellent quality. The Chittidar is similar to Safeda except that it is sweeter and has many pinkish red dots of the size of a pinhead on the surface of the fruit. Lucknow 49 is a seedling selection of Safeda, with a larger tree and larger fruit with rough surface. The Hafsi is a red-fleshed type. In Bihar, apart from Safeda, the other varieties grown are Harijha and Habshi with dark green fruit. Some of the other varieties are Dholka, Nasik and Dharwar.

The famous Allahabad Safeda as well as the Apple coloured Guava varieties have originated as chance seedlings in the village Abubakkarpur (now Bhabhakkarpur), Allahabad. Nand *et al.* (1991) reported ''Allahabad Surkha'' from the Allahabad guava belt. A four year old seedling was spotted in the village Sulem Sarai bearing fruits with deep pink flesh and uniformly pink skin. Organoleptic tests revealed that the fruits are sweet, strongly flavoured with moderate number of seeds. Daulta *et al.* (1998) reported new hybrids of guava. Guava Hisar safeda (H 25-35, a cross between Allahabad Safeda x Seedless has upright growth with a compat crown. Its fruits are round, weighing about 92 g each, pulp is creamy white with less seeds which are soft. TSS is 13.4% and ascorbic acid 185mg/100g. Hisar Surkha (H12-34) is a cross Apple colour × Banarsi Surkha. Of the pink fleshed hybrids it is ideal. Maximum fruit weight is found in L 49 (107g) followed by Hisar Safeda (92 g) and Allahabad Safeda (87g).

The guava generally has a short shelf-life which, however, varies with the cultivar. Several methods of pre and post harvest handling and storage strategies have been tested. Among the pre-harvest treatments, spraying of fruits with such chemicals as calcium nitrate, or a mixture of calcium nitrate and zink chloride, or with cycocel, appears beneficial (Singh and Chauhan 1993). The best treatmentregime is double spraying with a mixture of 1% calcium nitrate and 0.2% zinc chloride ; the treated fruit has reduced respiration as well as reduced firmness and weight loss, and an improved shelf-life. Proper handling prior to storage is important to preserve the fruit quality. Positioning of guava fruits in natural posture with the pedicel end vertically upward may result in a better keeping quaiity as compared with fruits kept in reverse position. The fruit seems to have a lower weight loss and a lower rate of ethylene and CO_2 evolution, higher soluble solids and ascorbic acid levels. Newspaper is frequently used as the cushioning material for the fruit. It is found that the guava cultivar L-49 individually wrapped, or packed in newspaper impregnated with biphenyl and/ or $KMnO_4$, and kept in room temperature showed less decay and retained a higher level of ascorbic acid than the control (Sharma *et al.* 1992). Biphenyl (0.5 g per pack) is more effective in reducing fruit decay than $KMnO_4$(3g per lit.), and individually wrapped fruits experience less water loss and decay than the packed fruits.

The guava has the potential to become commercially important tropical fruit. Some of the major constraints that might have limited the potential of the guava as an important export crop are problems related to pests and diseases, the availability of good dessert cultivar and the effectiveness of postharvest handling and storage technology.

Singh and Bhargava (1976) reported the presence of some antifungal substances in the skin of guava fruits. Of the three varieties taken, fruit skin of Chittidar and Apple guava hindered the spore germination of *Gloeosporium psidii* Delacr, *Pestalotia psidii* Pat. and *Phoma psidii* P. Henn. This result was more pronounced in case of Apple-guava variety where complete inhibition of spore germination of *Gloeosporium psidii* was observed even after 24 hrs of incubation.

3.2. BIRDS AND INSECT PESTS

Guava fruit is highly perishable and the birds play havoc with the fruit when it is not ripe. So it does not pay to plant only a few guava trees, since the cost of bird-scaring over a long period is prohibitive. Malhi and Kaur (2001) found birds damage in guava orchards was caused by rose-ringed parakeet (*Psittacula krameri*), followed by house crow (*Corvus splendens*), common myna (*Acridotheres tristis*) and house sparrow (*Passer domesticus*).

Guava suffers from a number of insect pests, a few important ones are as follows :

3.2.1. Oriental Fruit Fly : (Chaetodacus) Dacus dorsalis Hendel

(Family - Tephritidae : Order - Diptera)

Fruit flies damage the semi-ripe fruit in rainy season. The insects lay their eggs under the skin of the fruit. The maggots feed on the pulp of fruit and cause brown rotten patches thus making them unfit for sale.

Control

The affected fruit should be destroyed, the soil ploughed upto several times and the rainy season crop avoided. Three rounds of malathion 0.1 per cent in combination with either DDT (0.1%), Endrin (0.02%) or Fenthion (0.1%) at monthly intervals commencing $1\frac{1}{2}$ months after flowering, are recommended.

3.2.2. Bark Borer : Indarbela tetraonis Moore, I.quadrinotata Wlk.

(Family - Metarbelidae - Order - Lepidoptera)

These are confined to tropical region. Larvae are wood borers. Moths are nocturnal. The bark borer *I.tetraonis* is known to attack a number of trees such as mango, jamun, ber, orange, citrus, aonla, litchi, *Casuarina equisetifolia*, *Cassia* sp., etc. *I.quadrinotata* also occurs with this.

The larve bores into the stem to about 15 to 25 cm and feeds on the bark mostly during night under shelter of webs covering a considerable part of the stem extending from the bore hole. Only one larva is seen inside a bore hole which is used only as a shelter when it does not feed during day time. There may be about 15 to 30 larvae on a single tree. They enter through the bark into the wood and in cases of severe infestation sap movement in interfered with and the tree ceases to flush. Dr. Bhargava and his two associates, Dr. Arya and Mr. Khati, observed partially wilted plants in Babakarpur and Mundera in Allahabad. They helped in sprouting many dead trees in rainy season by cutting dead branches 30 cm above the soil.

The incidence of the pest is easily made out by the presence of the peculiar winding galleries, generally near the forks or angles on the stem and branches. Larva pupates inside the bore hole and the adult emerges leaving a portion of the empty pupal case protruding out of the bore hole.

Control

Injection of 5 ml of kerosene, CS_2, endrin 0.04 per cent or BHC (0.2%) or DDT (0.5%) into the hole and sealing it with cotton wool and mud gives good control. It is also preferable to spray the insecticide on the tree trunk after removing the webs so that effective control can be obtained.

3.2.3. Scale Insect : Pulvinaria psidii Gr.

(Family-Coccidae, Order-Hemiptera)

P.psidii is one of the commonest scale insects in South India. Besides guava, it infects mango, *Morinda*, tea, coffee etc.

Control

Spraying (0.05%) diazinon, parathion or malathion.

3.2.4. The White Grub : Holotrichia insularis Brenske

(Family - Melolonthidae, Order - Coleoptera)

This is the most injurious species of White grub pests noticed in Rajasthan. The Larvae feed on roots of plants in the seedling stage and plants dry up. In cases of heavy attack the plants wither and gradually dry up. A grub attacks a number of plants. The adults feed at night on the foliage of trees found in the vicinity.

The beetles emerge in June after the break of the monsoon. They are active at night feeding on the leaves of trees and defoliate them. The common trees defoliated by the adults are *Acacia arabica, Azadirachta indica, Carrisa carandus, Crataeva religiosa, Grewia asiatica, Mangifera indica, Punica granatum, Syzygium jambolanum, Tamarindus indica* and *Zizyphus jujuba* etc.

3.2.5. The litchi borer : Aristobia testuda Voet.

(Family - Cerambycidae, Order - Coleoptera)

The litchi stem borer of China is posing havoc to guava growers in the North-Eastern part of India. The plants are infested at certain localities, leading to complete death of the plants. If it spreads to the major guava-growing tracts of India, the pest will cause considerable economic damage to the growers. Hence, to check the spread and adoption of proper quarantine, measures should be taken up in the movement of guava saplings from the NEH region to the other parts of India (Shylesha *et al.* 1998)

The adults and grubs damage the trees as the grubs feed voraciously within the stem, making tunnels/gallaries up to 1.5—2 m long ultimately killing the tees within a year of attack. Adults of stem-borer, emerging from the pupae of previous year's infested plants is the primary source of infestation. Adults emerge by making circular holes of 1-1.5 cm diameter on the stem and after mating, lay single eggs on the branches and trunk of guava plant. After hatching, the grubs start feeding on the bark and later enter into the wood, 2nd cause considerable damage.

Control

The adults of trunk borer are conspicuous beetles, they can be collected during first week of June to first week of September. The plant is shaken and the beetles can be hand picked. Eggs laid on the plants can be clearly visible due to inverted U-shaped marking on the stem, and these can be destroyed by using a sharp knife.

Under severe infestation injection of 5 ml of Dichlorovos (0.1%) or Monocrotophos (0.2%) / hole gives effective control to the pest. On young plants (6-7 years old) application of Carbofuran granules @ 1kg / ha incorporated into the root zone also provides good control.

Soil application of insecticides and destruction of beetles by spray applications of parathion (0.05%) and endrin (0.1%) during their breeding season have been suggested.

3.2.6. Table : Some Other Insect Pests of Guava

Sr.No.	Name	Family	Order
1.	*Aphis gossypii Glov.(cotton aphid)*	Aphidiae	Hemiptera
2.	*Drosicha stebbingi Gr.*	Margarodidae	"
3.	*Ferrissa virgata Ckll.*	Pseudococcidae	"
4.	*Saissetia hemisphaerica Targ.*	Coccidae	"
5.	*S.nigra Nietn.*	"	"
6.	*S.oleae Ber*	"	"
7.	*Ceroplastes floridensis Comst.*	"	"
8.	*Aonidiella orientalis Newst*	Diaspididae	"
9.	*A.distructor Sign*	"	"
10.	*Helopltis antonii S.(Tea mosquito bug)*	Miridae	"
11.	*Rapala varuna Horsfield(Fruit borer)*	Lycaenidae	Lepidoptera
12.	*Achaea janata Linn.(Castor Semilooper)*	Noctuidae	"
13.	*Myllocerus discolor var.variegatus B. (Leaf weevil)*	Curculionidae	Coleoptera

DISEASES CAUSED BY FUNGI

3.3. SEEDLING BLIGHT

The disease caused by *Rhizoctonia* sp. (Sohi, 1983) manifests itself on seedlings upto 4 months of age. It is rarely observed in dry weather. The disease appears on leaves as small, circular to irregular brownish spots, which spread rapidly under humid conditions, covering the entire lamina and petiole. The upper leaves are the first to show the symptoms and the disease progresses from top downwards. The leaves drop-off and finally the whole plant dies.

Control

Spraying with Capton (0.2%) or Ferbam (0.3%) effectively controls the disease.

Cereospora sawada Yamamoto and *Phomopsis psidii* Nagraj & Ponnappa also attack the seedlings or young plants producing dark brown angular or irregular spots resulting in leaf culing. Spraying with Dithane Z-78 is beneficial (Sohi, 1983).

3.4. WILT OF GUAVA

It is one of the most destructive diseases in Uttar Pradesh and West Bengal. It results into splitting of the stem bark in early stages and ultimately results into yellowing and shedding of leaves. Within a few months of the attack the plant dries up. The disease was first noticed in 1935 in Allahabad (U.P.) and it spread to more than 20,000 sq. miles by 1952. It is estimated that in badly infected districts of Allahabad, Farukhabad, Fatehgarh, Lucknow, Faizabad & Unnao, 5 to 20% trees are killed annually (Tandon, 1967). The author observed 20-60% loss in some areas of Fatehpur and Allahabad district in 1981-82. According to Mehta (1951), the disease seems to occur on soil with a pH above 7.5. It becomes move severe with further increase of pH upto 9.0.

Fig. 21. Wilt of guava

Different pathogens have been reported to be responsible from different regions, depending upon the soil type and climatic conditions. In West Bengal under heavy moisture conditions it is caused by *Fusarium solani* (Mart.) App. & Wollenw and *Cephalosporium* sp., whereas in dry soil it is associated with *Macrophomina phaseoli* (Maubl) Ashby (*Rhizoctonia bataticola*). In areas where the soil moisture is moderate the wilt is caused by *Fusarium oxysporum f. psidii* (Prasad *et al.,* 1952, Sohi 1983). The symptoms produced do not vary to a great extent with the different pathogens involved. They further observed that the fungus could be carried from season to season on a mummified fruit.

Pandey and Dwivedi (1985) reported this organism responsible for wilt of guava in Varanasi district. They conducted pathogenicity tests and found that infected plants developed chlorosis followed by wilting of entire seedlings and leaf asscission. Histopathological studies revealed hyphae in xylem vessels of infected roots. Isolations from these roots gave positive results.

Dwivedi *et al.* (1990) found association of *Fusarium oxysporum* f. sp. *Psidii* and *F. solani* with the roots of guava. These were primary colonizers and *Macrophomina phaseolina* and *Rhizoctonia solani* as secondary associates. It was also found that built up of high inoculum potential of the pathogens causes pathogenesis.

Control

Pruning and burning of diseased twigs have often been suggested for the control of disease. The wilted plants should be uprooted and the stubbles dug out atleast to a depth of 1-1.2 metres and the pits should be disinfected by burning dry leaves. Soil may be treated with lime or gypsum (at a rate of 1 kg/tree) after exposing the roots to a depth of 15 cm. Resistant varieties like Banarasi (Andhra strain), Dholka a Sind, Nasik white guava, Supreme and Lucknow 49 should be planted. Use of resistant root stock and use of seedless gootee as root stock in the inarching method of propagation have also been found satisfactory against the wilt attack.

3.5. ASPERGILLUS ROT

Aspergillus rot usually starts as a water soaked lesion with brownish colour. It enlarges with age and gets depressed in centre. In later stages black mouldy growth of *Aspergillus niger* appears on the infected surface of the fruit. The whole fruit becomes infected with black conidial heads which develop well between 10 to 12 days, Lal *et al.* (1980) reported 20-25% loss of the fruit during harvest and transit. Seven species namely *A.niger, A.nanus, A.parasiticus, A.phoenicis, A.flavis, A.flavips* and *A.awamori* are reported on ripe guava fruit. Lal *et al.,* (1980) reported *A.awamori* Nakazawa from Allahabad. They observed that high humidity and high temperature 30-35°C favoured the development of disease. However, very young green fruit failed to show any soft rot symptoms.

Singh and Sinha (1983) reported biochemical changes in guava fruit by two species of *Aspergillus. i.e. A.flavus* Link and *A.parasiticus* Speare produced 0.563 and 0.257 ppm of aflatoxin B, in inoculated guava fruits respectively.

Control

Arya *et al.* (1981) tried four systemic fungicides viz., Delan, Saprol, Tecto-40 and Bavistin against *A.awamori*. They recommended the use of 1250 ppm of Bavistinand Saprol for 2 min as pre-inoculation treatment to control Aspergillus rot. Krishnaiah *et al.* (1985) tried post-harvest protection of guava fruits using Decco food-grade fruit coatings. They used the coating at 0.04 ml/100g fruits and the spoilage fungi included *Aspergillus, Botryodiplodia, Colletotrichum, Fusarium, Glomerella, Gleoesporium, Pestalotia, Penicillium* and *Phytophthora*.

3.6. STYLAR END ROT

Causal organism = *Phomopsis psidii* Nagraj and Ponnappa
P.destructum Rao, Agarwal & Saksena

The disease starts as circular water soaked lesions at the stylar end. With the advancement of rot, small, white or light brown, pycnidia appear on the infected region. They are often very closely aggregated and discharge masses of spore which appear as pinkish buff. Lim and Rajak (1986) reported rot in Kampuchean guava from Malaysia causing yield losses more than 50%.

Its occurrence was first reported from Lucknow by Rai (1956). The disease was subsequently identified by Srivastava and Tandon (1969). Both die back and fruit rot caused by *P.psidii* are reported from lower Kumaun (Uttranchal). The percentage rot was 85% in the 'apple guava' as compared to 100% in 'Allahabad Safeda' after 15 days of incubation (Arya, 1982). Rao *et al.* (1976) reported it from Sagar and found the causal organism as *P.destructum*. In Sept. Oct. 1995-96 a severe out-break of stylar end rot caused by *Phomopsis psidii* was observed in Fruit Research station Kothiyalseh, Chamoli in Garhwal Himalayas, Uttranchal (Upadhyay and Tripathi 1996).

Lal *et al.* (1985) reported decrease in α-alanine, glutamic acid, aspartic acid, serine-glycine, asparagine and some of them like leucine-isoleucine, valine, β-alanine and histidine-lysine were completely utilized from diseased tissues and hence could not be detected chromatographically. During pathogenesis, a regular increase in organic acids was observed upto 10th day. The quantity decreased subsequently. Tartaric and malic acids were altogether absent after 15 days. Since appreciable amount of malic acid was present in 15 days old healthy guava fruit, it was presumed that *P.psidii* utilized this organic acid from the infected tissue. The absence of sucrose and the increase in glucose was recorded in 5 days infected fruit. Only glucose was left in the infected tissues after 15 days of incubation.

The percentage loss of ascorbic acid in healthy guava fruits was 19, 40.4 and 8.6% while it was 67.2, 79 and 98.4% respectively in 5, 10 and 15 days infected fruits.

Control

Arya (1988b) tried leaf extracts of *Azadirachta, Eucalyptus, Ocimum* and *Strychnos* in controlling stylar-end rot of guava. He found post-inoculation treatment with *Strychnos* and *Eucalyptus* more effective. Arya (1988a) found satisfactory control with Bavistin, Difolatan, Calyxin (100 ppm). Dithane M-45 (1250 ppm) could control 62.5% disease. Leaf extracts of *Eucalyptus occidentalis* Skeels and *Aegle marmelos* Corr. (50%) were effective. Amongst antibiotics and fixed oils, Idifulvin & mustard oil were most effective. Rao and Agarwal (1976) suggested the use of Bavistin in controlling the disease caused by *P.destructum*.

3.7. GREY BLIGHT AND FRUIT CANKER

Causal Organisms — *Pestalotia psidii* Pat. (Tandon, 1967)
Pestalotiopsis clavispora (Atk.) Stey.
(Jamaluddin and Tandon, 1977) and
Pestalotiopsis palmarum (Cooke) Stey.

The pathogens attack leaves and fruits. On the leaves, reddish brownish angular to irregular spots are formed, on the fruit the disease starts in the form of a brownish discolouration which attains a russet colour after a week. With the enlargement of he spots their central region gets slightly depressed. Subsequently black minute single or gregarious acervuli with viscid spore masses appear on the infected tissue. In partially rotton fruits the margin of the infected tissue appears ochraceous buff in colour. Young diseased fruits do not attain normal size, get cracked and are mis-shapen.

Fig. 22. Development of disease caused by *A.awamori*.

1. Healthy fruit, 2. Fruit showing distinct ring, 3. Eight days infected fruit.

Pathological studies revealed that *P.psidii* hydrolysed the entire sucrose content of the guava fruit within six days. Presence of valine in fruits infected with *P.psidii* was observed while it was absent in all other diseases, including healthy fruits. Effect of temperature on the guava fruit revealed that there was no rot at 10°C, but after incubation at this temperature the rate of decay was rapid when the fruits were bought back to room temperature (25° ± 2°C). When the fruit was kept at 35°C the disease retarded but at this temperature the fruit shrivelled up, its pulp became brownish and it was rendered unsuitable for consumption. Most diseases occurred at 25-35°C and high RH (Kaushik *et al.,* 1972). Khanna and Chandra (1976) reported control of *P.psidii* by dipping guava fruit in 50 ppm aratan for 2 minutes either before or after inoculation. Control of Pestalotia rot by leaf extracts of certain medicinal plants was tried by Pandey *et al.* (1983), 75% extract of Neem and Tulsi controlled

100% and 90% disease as pre-inoculation treatment, it was 80% and 70% respectively in post-inoculation treatment. Bhargava & Singh (1974) reported 80% control with 2000 ppm of Tecto-40.

Control of guava fruit rot by Homeopathic drugs was tried by Khanna and Chandra (1989). Castor oil used as adjuvant improved the efficacy of *Kali iodatum* potency 87. At this concentration 75% fruits were saved from disease.

Guava fruits treated with Kali iodide 87 and Arsenicum album 181 before inoculation failed to develop any rotting. It was observed that post-inoculation treatment with these drugs was not much effective (Khanna & Chandra, 1977).

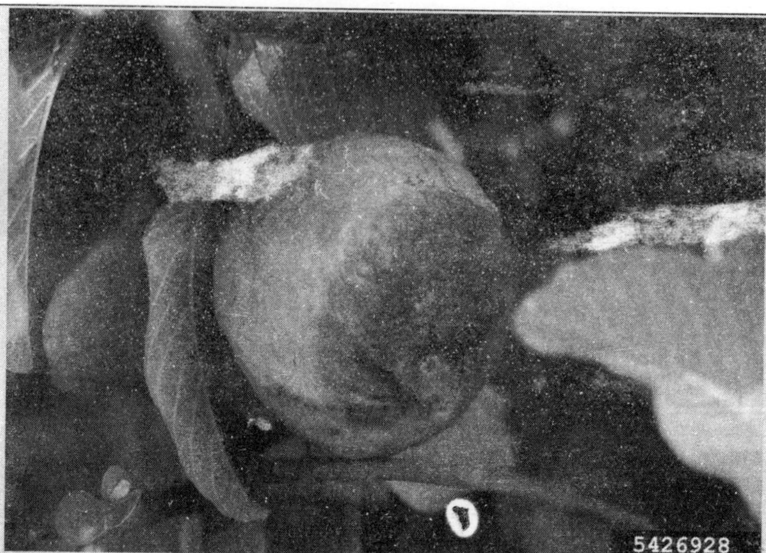

Fig. 23. Guava fruit showing natural rot of *A.awamori*..

3.8. ANTHRACNOSE

Guava anthracnose is caused by four pathogens :

Gloeosporium psidii Delacr, *Colletotrichum gloeosporioides* Penz. (*Glomerella cingulata*)

Colletotrichum acutatum Simmonds and *C.psidii Curz.*

In Assam 90% of the guava fruit rotted within 5-10 days of storage the pathogen was reported as *Colletotrichum acutatum* by Das and Bora (1993). On the fruits fungus formed acervuli. The conidial ooze was seen on ripe rots under humid conditions. Five isolates of *Colletotrichum gloeosporioides* (*Glomerella cingulata*) were grown on PDA, Czapek's yeast agar and guava docoction agar (Wahid 1999). Their sensitivity towards Rovral (iprodione) and Sumilex (procymidone) was tested. Some fungi isolated from guava fruit surface could inhibit the growth of *G.cingulata*.

Guava fruits were infected by various fungal pathogens during storage in Assam resulting in complete rotting of the fruits within 7-10 days. The fungal pathogen found responsible was

C.acutatum. The infection process was studied using lightmicroscopy and electron microscopy. Infection hyphae grew directly from conidia and entered the fruit wall through stoma or directly penetrating the cuticle. Subsequent intercellular and intracellular development of the fungus resulted in the rapid degradation of cell wall and membrane's structural integrity leading to the rapid degradation of cell wall and membrane's structural integrity. It resulted in collapse of cells and ultimately forming acervuli which pierced the fruit cuticle (Das and Bora, 1998).

Many small, shallow, water soaked lesions are produced over the surface of the fruit. With age they enlarge and get depressed. Subsequently they coalesce to form bigger spots of irregular shape and size and under humid conditions they develop salmon coloured spore masses at the centre of such lesions. The causal organism has been identified as *Gloeosporium psidii* Delacr.

Collectotrichum psidii Curz. causes dark brown spots on leaves and fruits. It spreads very fast on leaves covering most of the leaf area. On the fruit, the spots remain small and circular. The pulp beneath the spots becomes hard. It causes fruit rot in storage.

Control

Spraying of trees with 0.2% Difolaton or Dithan Z-78 is beneficial. Tandon and Singh (1969) obtained good control by spraying with Bordeaux mixture 3 : 3 : 50 + Copper oxychloride and Cuprous oxide at weekly intervals. 500 ppm of Tecto-40 could control 40% of fruits on 12th day due to infection by *G.psidii* (Bhargava & Singh, 1974). Protection against *C.gloeosporioides* was given by dipping guava fruit in 50 ppm aratan for 2 minutes either before or after inoculation (Khanna & Chandra, 1976).

Gamma radiation (Ionizing radiations) were used for post-harvest decay of Guava caused by *C.gloeosporioides* (Gupta & Chatrath, 1973). There was progressive decrease in the germination of the conidia of *C.gloeosporioides* and length of the germ tube with increase in the doses of gamma rays above 10 Kr., irradiation of fruits with gamma rays at a dose of 100 Kr. after 5, 18 and 24 hr of inoculation gave good control of the disease while fruits were completely rotten in control and in treatment which was given 42 hr after inoculation. Pre-inoculation irradiation treatment with gamma rays at the same dose was not effective.

3.9. THIELAVIOPSIS ROT OF GUAVA

Lal *et al.* (1980) reported a new soft rot of guava caused by *Thielaviopsis* state of *Ceratocystis paradoxa* (Dade) C.Moreau, which was responsible for extensive damage to the harvested fruits in the markets of Allahabad and neighbouring districts. The author has found this disease common in Baroda. The fungus has been earlier reported from arecanut, coconut, custard apple, papaya and tomato.

Artificial inoculation of the fungus on freshly harvested, healthy, firm and nearly mature fruits of guava resulted into a circular water soaked area with white mycelial growth in the centre which turns black after 5 days due to the production of large number of endoconidia and black chlamydospore. Thus a black crust composed of dark coloured fungal mycelia and spores encircles the whole fruit within 8-10 days of incubation. Finally the fruit loses its firm texture, gets deformed and the flesh of the fruit turns black, soft and pulpy. The diseased fruits emit a foul odour.

Control

To control the thielaviopsis rot of guava use of 500 ppm of Bavistin, 1000 ppm of Difolatan as post-inoculation treatment and 500 ppm Benlate as pre-inoculation treatment is recommended (Lal *et al.,* 1982). They also recommended to immediately discard the diseased fruit and avoid any injury on the flesh during harvest, transit and marketing.

3.10. MYROTHECIUM ROT

Rai *et al.* (1982a) reported soft rot of guava caused by *Myrothecium roridum* Tode ex Fr. The disease was observed in Aug.-Sept. The disease started as slightly sunken, small, nearly circular, olive grey patch which gradually increased in size and was covered with enormous growth of white woolly mycelium with dark green viscid spore mass. The pathogen remained restricted to the infected region.

3.11. PHYSALOSPORA ROT

During the course of study of fungal fruit rot diseases of guava in the vicinity of Allhabad Rai *et al.* (1982 b), reported a new storage rot caused by conidial stage of *Physalospora psidii* Steven. & Peirce. Earlier the fungus has been known to cause leaf spot disease in guava (Stevenson and Peirce, 1933).

The disease starts from stylar end of fruit. The symptoms start in the form of circular brown coloured water-soaked lesions changing to dark brown spots which gradually advance towards stem end and covered the whole fruit within 7-8 days.

3.12. MACROPHOMA ROT

Kapoor and Tandon (1970) reported Macrophoma rot of guava from Allahabad caused by *M.allahabadensis* Kapoor & Tandon.

In the early stages the diseased fruit showed brownish and water soaked discolouration of the skin spreading around the point of infection. The mycelium on the surface of the fruit was orange,. Cinnamon to dark ivy-green in the beginning, but eventually it turned dark brown to black and covered the whole fruit. In later stages numerous dark coloured pycnidia of *M.allahabadensis* appeared all over the surface. No other species of *Macrophoma* has been found associated to guava fruit.

Control

Twenty fungicides were evaluated in laboratory against *Macrophoma* sp. (Quintero and Urdaneta, 1998). Best control was achieved with Fosetyl-aluminium (as Aliette). Other effective chemicals were Carbendazim (as Bavistin) and Copper oxychloride (as Cobox).

3.13. RHIZOPUS ROT

The lesions are usually water soaked and circular in outline. As the decay progresses the affected area becomes covered with strands of coarse mycelial growth. The decay caused by the pathogen is soft and watery in nature. The causal organismis *Rhizopus nigricans* Ehrenberg (Tandon, 1967).

3.14. BELTRANIA ROT

Rai (1982) found *Beltrania rhombica* Penzig on ripe fruits. The disease startsas water soaked spot developing black in colour after 6-7 days due to production of dark coloured conidia.

3.15. PHOMA ROT

Brown coloured circular spot (2.5 to 3 cm in diameter) is produced over the surface of the fruit. The centre gradually gets depressed and the water soaked peripheral margin remains slightly elevated with sparse fungal growth. Minute dot like pycnidia appear and they ooze out cream coloured pycnidiospores. The decay causing organism has been identified as *Phoma psidii* P.Henn. (Tandon, 1967).

Control

1000 ppm of Tecto-40 controlled 60% disease on 12th day (Bhargava & Singh, 1974).

3.16. BOTRYODIPLODIA ROT

The infection initiates as brownish discoloration at the stem end and gradually proceeds downwards in an irregular wavy manner. Finally the whole fruit may get involved. In advanced stages numerous small pycnidia are produced throughout the surface of the fruit. The rot produced by the pathogen is soft and watery in nature. The disease is caused by *Botryodiplodia theobromae* Pat.

Patel and Pathak (1993) found presence of numerous fungi in markets of Udaipur. *Rhizopus* and *Botryodiplodia* were main pathogens. Monthly losses were calculated in Ahmedabad and Udaipur markets. Loss was 2.68% in Ahmedabad and 2.8% in Udaipur due to this disease alone.

3.17. PHYTOPHTHORA ROT

At least three species of *Phytophthora* are described to cause this rot in different tropical areas of the world. *Phytophthora citricola* causes a fruit rot of guava in Hawaii. Fruit develop a greyish brown colour ; mature yellow fruit are less susceptible to the disease than are mature green fruit (Ko *et al.*, 1982). Thankamma (1983) reported *P.heveae* causing fruit rot of guava. Isolates of this fungus from guava did not affect rubber.

Phytophthora parasitica is pathogenic in India and Cuba. This fruit rot is the most serious disease of guava in Karnataka during the rainy season. It is also a problem in Tamil Nadu, Andhra Pradesh and Maharashtra states, and to some extent in Rajasthan. Singh et al.(1978) successfully isolated *P.parasitica* from the soil beneath guava trees in India and suggested that the soil may be main source of the infection.

The disease appears in rainy season, fungus attacks unripe fruits at the blossom end. It produces small conspicuous dark brown circular spots, these spots may cover the entire fruit in 3-4 days of humid weather. Diseased fruit become somewhat softer and have an unpleasant odour. Some of the diseased fruits may be shed prematurely. The incipient infection on fruits develops further on storage and results in appreciable damage. Sohi and Sridhar (1971) reported that the disease resulted in a loss of 20 to 36 kg in yield per tree at Hessaraghatta during 1969 and 1970 ; the total loss per acre was estimated at 2160 kg and 3888 kg respectively. Fruit rot of guava has also been a significant problem in the Punjab, with disease incidence ranging from 5 to 20%. Affected varieties in the region are Allahabad Safeda, Apple guava, Red Fleshed, and Pink Fleshed (Singh *et al.,* 1976).

Control

Sharma *et al.* (1978) obtained good control of Phytophthora rot with Bavistin (Carbendazim). Hot water treatment was not effective.

3.18. FUSARIUM ROT

Severe losses were reported by *Fusarium solani* (Mart.) App. & Wollenwin in the markets of Burdwan and its neighbourhood (Chakrabarti, 1983). Rai isolated it from the fruits in Allabahad (Rai, 1982).

3.19. BOTRYOSPHAERIA ROT

Brown rot on guava fruits caused by *Dothiorella* conidial phase of *Botryosphaeria dothidia* was observed in Venezuela. The organism was grown in PDA and oatmeal agar medium. 2 isolates produced multiloculated conidiomata with holloblastic phialides and colonies that initially were white and later on turned to grey and eventually black (Cedeno *et al.,* 1998).

3.20. TABLE : SOME OTHER DISEASES ON GUAVA TREE

Parts used for isolation	Organisms	Locality	References
Roots and Soil	*Fusarium oxysporum f.sp.psidii*	Lucknow	Tandon, 1967
	Prasad, Mehta & Lal	Allahabad, Varanasi	Pandey and Dwivedi, 1985
Leaf	*Curvularia lunata (Wakker) Boedijn*	Allahabad	Tandon, 1967
	Botryodiplodia theobromae Pat	Allahabad, Jaipur, Chennai	Tandon, 1967

Fruit and Flower	***Nigrospora oryzae* (Berk & Br.) Petch**	Allahabad	Tandon, 1967
	Pestalotia psidii Pat	**Allahabad,**	Tandon, 1967
	Phyllosticta sp.	**Jaipur, Chennai**	
	Gloeosporium psidii	**Pratap Garh,**	Tandon, 1967
	Delacr.	**Jodhpur**	
	Sordaria sp.	**Allahabad**	**Tandon, 1967**
	Gloeosporium psidii	**Allahabad**	
	Delacr. mummified fruit	**Pratap Garh**	
	Chaetomium globosum	Allahabad	Rai, 1982
	Kunze & Schm.	Allahabad	Rai, 1982
	Alternaria alternata (Fr.) Keissler	Chennai	

3.21. ZINC-DEFICIENCY DISEASE

This is common in sandy soils (Rajasthan) or alkaline soils (Delhi). The symptomsof the disease are a severe reduction in leaf size, interveinal chlorosis, growth suppression, dieback of shoots and cracking of fruits, resulting in a heavy reduction of yields and quality.

Control

Carry out foliar spraying with a dose of Zinc sulphate and hydrated lime 0.7 lb in 16 gallons of water : also soil application of Zinc sulphate at 4 to 12 oz. Per plant (Ray Chaudhuri and Lele, 1967).

4

Banana

4.1. GENERAL INTRODUCTION

The banana is one of the most ancient of food plants having been used and perhaps cultivated at the dawn of civilization. It is one of the most common plants in a large part of the world and its production is more highly industrialised than that of any other fruit. Hays (1960) has quoted Disraeli, according to whom, the most delicious thing in the world is banana.

Banana is known as 'Adam's fig', 'Apple of paradise', 'Kela' and botanically *Musa paradisiaca* L. which belongs to family Musaceae. The fruit is available throughout the year and, being relatively less expensive compared to other fruits, can be called as a fruit of the masses.

The edible banana is indigenous to the warm moist parts of Asia and probably originated somewhere in mountainous regions of Assam, Burma, Thailand or Indo-China. From here it has spread to tropical parts of America, Africa, Australia, Philippines and Hawaii. Now it is industrialised in tropical America, Hondruas, Jamaica, Brazil and Mexico who are the biggest exporters of banana.

Unlike most other fruits, banana is a herbaceous plant. Banana bears enormous leaves on a pseudostem consisting of clasping leaf-stalks. The stem is a greatly reduced underground corm, from which the pseudostem arises. It produces two sets of roots one growing down upto a depth of about 2 metres and other spreading horizontally. The flowering stalk pushes out of the pseudostem and produces a large hanging bunch of flowers. The flowers are borne in the exils of brightly coloured bracts called spathe. The basal flowers are female flowers, then appear the male flowers, followed by neuter flowers towards the tip. Pollination is not required for fruit set. The pseudostem produces flowers only once and is cut off after fruiting.

The fruits are called fingers, which are borne in hands of about 12 fingers each on the bunch. Usually a bunch has upto 150 fingers.

The fruit is delicious and seedless berry. Soft ripe banana is the first solid food given to babies. Almost all parts of the plants are used. Fruits are taken as dessert or made into custard and fruit cream. Unripe banana fruit as well as the inner core of the pseudostem is regularly cooked as a curry in the banana growing areas. The end of the inflorescence is cooked as a vegetable in Bengal.Ripe bananas are dried into banana figs. Unripe banana fruit is used for making salted banana chips. Banana pulp is also dried into flour. Chopped banana stems are used as cattle feed. Dr. Amanat Ali reported from Pakistan that banana plants after fruiting are cut and mixed with sugar molasses and wheat straw to make the Silage which is used as green fodder for animals (Arya, 1990). Apart from being used as a food and fodder, the banana has another equally important use. Its leaf is universally used as a plate for serving meals in various parts of India, especially in the South. The use of metal plates for this purpose is a practice which started in modern times. In Tanjore district in Tamilnadu alone, 2000 acres of banana are reported to be cultivated exclusively for leaves. Some species of banana also yield fibre. *Musa textilis* forms the basis of a regular industry for making Manila hemp in the Philipines and *M.basjoo* is used for making coarse fabrics in Japan (Ranjit Singh, 1969).

The banana is nutritious, easily digested fruit, rich in carbohydrates and minerals, such as potassium, magnesium, sodium and phosphorus, and is even richer in calorific value than potato. The ripe fruit is a good source of Vitamin A and a fair source of Vitamin C, B_1 and B_2. The unripe fruit is even richer in Vitamins which are retained well even after cooking at a temperature upto 60°C. The fruit is a fair source of calcium and iron. The ripe fruit contains upto 27 per cent sugars, which consist of both the reducing and non-reducing types.

In India the banana is one of the most important fruits ranking next only to mango in area and production. The total area under banana in India is 4,64,000 acres (Ranjit Singh, 1969). It comprises 15 per cent of the area under fruits. The banana cultivation is concentrated mostly in Tamilnadu, Maharashtra, Kerala, West Bengal, Andhra Pradesh and Karnataka. Most of the area in Assam and Bengal consists of small backyard plantations.

Poovan is an important variety and is extensively grown in old Madras state, Assam, Travancore-Cochin, Mumbai and elsewhere, except the interior areas of Maharashtra. Some of its other names include Chakrakeli, Lal Velchi, Champa and Chini Champa.

The second important variety Basrai has still longer list of synonyms. It is a dwarf variety and is grown under various names including Mauritius, Vamankeli, Kabuli, Sehafi, Bhusaval, Pacha Vazhai, Jahaji, Hirvi and Bengali.

Mortaman (Rasthali, Mutheli or Sonkel) is similar to that of Poovan in appearance, but of a much better quality. Some semi-tall varieties are regarded by Nayal and Bakthvathsalu (1955) to be their mutants and they are called Harichhal or Bombay green, Pedda Pacha, Arati of Andhra, Malbhog of Bihar and Sapri of Assam.

Nendran (Rajeli) is a dual purpose variety of great importance in Kerala. The fingers are long, angular prominently tapering and retain some starch even on ripening.

Monthan in a variety which can withstand some cold and is found growing in sheltered places in U.P. Apart from above, there are many more varieties grown in different parts of the country.

Banana cultivation can also be developed in terrai regions of Bihar and Uttar Pradesh, which are free from hazards of severe frost. While attempts have to be made to increase the area under banana, it is none the less important to improve production and handling techniques to obtain more yield, better quality and minimize losses in storage, transport and marketing, so as to ensure greater profits to growers from the existing plantations.

Like mango the banana is also harvested raw and ripened artificially. The fruit is harvested after the fingers begin to turn light green and the ridges on their surface change from angular to rounded. The top leaves at this stage begin to dry up. The entire bunch is cut with a one-foot-long stalk which acts as a handle. Smearing the cut ends with vaseline is recommended. The bunch should be protected from the Sun. Banana is commonly ripened in a closed room with smoke, but it can also be ripened at room temperature. Insertion of a little garlic into the cut end of the stalk of the bunch is said to hasten ripening. In a closed atmosphere the moist heat and the ethylene gas generated by the ripening fruit hasten the ripening of other fingers.

According to Van Loesecke (1949) the temperature of 22-30°C, relative humidityand supply of gases like ethylene, acetylene, ammonia and ozone. Sulphur dioxide used in some cases causes darkening and discolouration of skin without affecting the pulp. Smoke treatement for the duration varying from 8-24 hours is a common method of curing banana fruit in special rooms or store houses.

Bhan (1967) reported that experiments conducted at the Central Food Technological Research Institute (CFTRI), Mysore, have shown that the banana fruit can be stored at temperatures between 25 and 27°C with 85 to 90 per cent humidity for two to three weeks.

Banana, one of the popular crops in India, generates 850 lakh tonnes of banana waste per annum. The complete physio- biochemical characterization of this wastereveals abundance of various nutrients in it. So this waste is a suitable substrate for various biotechnological processes, among that mycoprotein production is the prominent one (Singh *et al.* 1999).

Survey of fruit markets in Gauhati has been done and aeromycoflora is studied by Sharma and Bhattacharjee (2000). They found presence of *Aspergillus, Alternaria, Botryodiplodia, Botrytis, Cercospora, Colletotrichum, Curvularia, Cladosporium, Fusarium, Helminthosporium, Mucor. Pencillium, Rhizopus, Trichoderma* and *Verticillium* in air. Baruah *et al.* (2000) studied fungal spoilage of banana in Eastern sub Himalayan region. They found dominance of *Asperagillus terreus, Fusarium oxysporum, F. moniliforme, Penicillium italicum, Tricohecium roseum* in this region.

4.2. INSECT PESTS

4.2.1. The banana aphid : Pentalonia nigronervosa Cog.

(Family—Aphididae : Order—Hemiptera)

The banana aphid is distributed in Australia, Sri Lanka and India. In India it is prevalent in Kerala, Tamilnadu, Karnataka, Maharashtra, West Bengal and Delhi.

It is a well known vector of the banana buchy top virus, Cardamom mosaic (''Katte'' disease) Amomum mosaic and 'Foorkey' disease of large Cardamom. The bunch top disease of banana was first noticed at Kottayam (Kerala) in 1943. Since then it has spread to various parts. It is also found on Cardamom in South India during July to Feruary.

To aphid occurs on banana usually in small to very large colonies consisting of alate and apterous viviparous females and nymphs. Apterous females are predominant in a population. On banana plants the aphids are seen at the base of the pseudostem, the top region of the pseudostem and the leaf exils. If abundant they may be found in colonies on an exposed surface of tender leaves, mostly on underside, on the sides of midrib or at the top and in unopened leaves. The alete and apterous females reproduce parthenogenetically and give birth to nymphs. No males are known. The nyphel period is 8 to 9 days. The longevity of an adult aphid is 23 to 27 days.

Control

1. Parathion and endrin are toxic
2. Spray of systemic insecticide like dimethoate, methyl demeton, thiometon and phosphamidon, etc. can be done to reduce the incidence of banana aphid on banana and Cardamomum.

4.2.2. Rhizome borer : Cosmopolites sordidus G. & Pseudostem borer : Odioporus longicollis Oliv.

(Family—Curculionidae : Order—Coleoptera)

The adult beetles are about 1 cm. long dull brown coloured. The head is more or less prolonged into a snout which considerably varies in different species. The main body is clothed with scales. When disturbed the adult beetles fall to the ground and remain motionless.

The two species of insects *C. sordidus* and *O. longicollis* which eat into the underground rhizome of banana from April to October, are the most serious pests in Kerala, Orissa, Karnataka and Tamil Nadu. The variety Malbhog is especially susceptible in water logged conditions.

Control

1. Aplication of a spray of aldrin (0.05%), DDT (0.16%) or BHC (0.05%)to the soil around the plant has been recommended.

2. Aerial spray of DDT has also been suggested. However, their effectiveness is limited since the grubs live inside the rhizome.
3. A practical remedy is to burn the infested rhizome and plant banana in a new well drained soil with healthy vigorously growing suckers.

4.2.3. Leaf thrips : Panchaetothrips indieus Bagn

(Family—Thripidae : Order—Thysanoptera)

This thrip also infects turmeric. Infestation causes rolling up of leaves, they turn pale yellow and ultimately dry. The rhizomes do not develop or are ill developed.

Control

Dusting or spray BHC, Carboryl (0.1%), and monocrotophos (0.03%) are effective.

4.2.4. White ant : Glyptotermes Dilatatus Gr.

(Family—Kalotermitidae : Order—Isoptera)

Termites or white ants are social insects, living together in organised communities composed of mature male and females and sterile workers and soldiers. They occur in the tropics and warm temperatrue regions.

The name isoptera refers to the two pairs of similar wings. Kalotermitidae is a group of primitive wood dwelling termites without workers but with pseudergates of pseudo workers. Pronotum flat, broader than head; wings without anal lobe.

They feed on the young rhizome of banana.

Control :

1. Incorporation of aldrin (5%) dust or heptachlor (5%) dust at 25 kg/ha in the soil before sowing has been recommended.
2. Soil application of BHC or post planting treatment applied in irrigation water affords protection.

4.2.5. Armoured scale : Aspidiotus cyanophylli Sign.

A. *cydonia*e Cockerell

A. *destructor* Sign.

A. *transparen*s Gr.

(Family—Diaspidiae : Order—Hemiptera)

The adult female is soft bodied, small and remains freely under a scale covering formed of wax secreted by its body. The scale of male is more elongate and smaller than that of the female. The adult female is flattened, disc like, apodous.

The insect reproduces bisexually or parthenogenetically. It is also a destructor of coconut.

Control

Spraying parathion or dimethoate may be useful.

4.2.6. Leaf bettle : Nodostoma subcostatum J.

(Family—Eumolphiae : Order—Coleoptera)

Leaf and fruit beetle *N. subcostatum* which chews the fruit and the leaves seriously from August to September, is also important from Assam to U.P. Certain varieties of banana like Alpan, Champa and Malbhog are especially susceptible.

The bettles are usually oblong, convex, either metallic or yellowish with spots. The beetles are characterised by prothorax narrower than elytra and legs not compressed.

Control

It can be controlled by clean cultivation and spraying of DDT (0.1%) or BHC.

4.2.7. Bag worm : Kophene cuprea M.

(Family—Psychidae : Order—Lepidoptera)

Bag worm moths are widely distributed insects. The females are apterous whereas males are winged. Wings are thinly clothed with imperfect scales and exceptionally large frenulum. Antennae are bipectinate some females are devoid of antennae, mouth parts and legs. Larvae inhabit cases which exhibit great variety of shape and of materials used in their construction.

The Bag-worm *Kophene cuprea* is a pest of banana in Kerala.

4.2.8. Flower thrips : Thrips hawaiiensis Morgan.

(Family—Thripidae : Order—Thysanoptera)

Leaves and flowers are attacked by *T. hawaiiensis*. It belongs to family Thripidae and subfamily Thripinae where Antennal segments 3 and 4 with slender sense cones, which are simple or forked. Antennae 6 to 9 segmented. Terminal antennel segments not long and thin dorsum of body not polygonally reticulate. Foretarsus sometimes with a claw-like appendage. Life cycle completes in 11 to 21 days.

Control

1. Dusting or spraying BHC
2. Spray applications of dimethoate (0.03%) has been found effective.

4.2.9. Fruit Fly : Daucus dorsalis Hendel

(Family—Tephritidae : Order—Diptera)

Fruit flies are medium sized insects usually with spots or bands on wings. These insects are of immense importance as the larvae of most species are phytophagous. It is important as it destroys the fleshy fruit of banana (for description of oriental fly *D. dorsalis,* see 1.2.2.)

4.2.10. Banana mealy bug

Pseudococcus comstocki (Hemiptera : Pseudococcidae)

This pest is often imported bunches of bananas, when groups of the white wax-covered insects can be seen on the base of the fingers and the bunch cushions. In the field insects can be found on all above ground parts.

Control

Phosdrin sprays are effective as are aldrin and dieldrin, but the main difficulty is in reaching the pests protected by leaf bases. Dichlorovos has also been suggested.

4.2.11. Red Spider Mite

Tetranychus lambi (Acarina : Tetranychidae)

The warm humid conditions under plastic bunch covers can be very favourable for a build up of red spiders. When such a built up occurs the red spiders damage the skin of the bananas and cause it to turn red. The damage is caused to the fruit on stalk end. Mites normally occur in colonies under side of the leaves (Feakin 1972).

Control

When Red spider occur 0.05%, spray of dicofol (2, 2, 2 trichloro- 1, 1, di(4 chlorophenyl) ethanol is suggested.

4.2.12. Table : Some other Insects Pests of Banana

Sr. No.	Name	Family	Order
1.	*Schistocerca gregaria Forsk,, (Desert locust)*	Acrididae	Orthoptera
2.	*Phymatostetha deschempsi L. (Spettle insect)*	Cercopiae	Hemiptera
3.	*Ferrisia virgata Ckll.*	Pseudococcidae	Hemiptera
4.	*Lecanium discrrepans Gr.*	Coccidae	Hemiptera
5.	*Aonidiella aurantii Mask.*	Diaspididate	Hemiptera
6.	*Prodromus subviridis D. (Mirid bug)*	Miridae	Hemiptera
7.	*Chaetanaphothrips signipennis Bagn (Fruit-rust thrips)*	Thripidae	Thysanoptera
8.	*Astrothrips parvilimbus Stan. and Mitri (Fruit peal thrips)*		
9.	*Helionothrips kadaliphilus R & M (Leaf thrips)*	Thripidae	Thysamoptera
10.	*Scolothrips asura R. & M.*	Thripidae	Thysanoptera
11.	*Latoia lepida C.*	Thripidae	Thysanoptera
12.	*Pericallia ricini Feb.(Wooly bear)*	Aretiidae	Thysanoptera

4.3. VIRUS DISEASES

4.3.1. Bunchy top

Among the virus diseases of banana, bunchy top is most important. It is prevalent in many countries such as Fiji, Pacific Islands, Australia, Borneo, Malaya, Sri Lanka, Egypt and

Belgian Congo. The disease was first recorded in 1879 from Fiji and in India, from Kerala in 1943. It slowly spread to Tamilnadu, Karnataka, Andhra, Maharashtra, Orissa, Bihar and Assam and probably in West Bengal and other states.

Fig. 24. Bunchy top of banana .

The disease is caused by Bunchy top virus, Banana Virus I or Musa Virus I. Bunchy top is spread in the plantations by the banana aphid (4.2.1.) when it sucks the virus-infected sap of the diseased plant and then leaves it and feeds on a healthy one.

The primary symptoms of bunchy top are seen if infected suckers are planted. They putforth short, narrow leaves which are chlorotic and exhibit mosaic symptoms. The leaves have distinctly up-curved margins. They fail to bend over normally and retain a stiff erect habit which, combined with the fact that the leaf stalk is reduced in length, gives the characteristic rosetted appearance to which the disease owes its name. The foliage on such plants is crisp and brittle when crushed. A bunch is rarely produced unless infection has taken place late in the life of the plant. The diseased plant neither dies nor recovers from the infection.

In early stage the youngest leaves exhibit a light green colour along the edge. A definite and unquestionable diagnosis can be made by examining the base of the youngest leaf from the underside and with the light behind it. If the plant is infected there will be narrow, dark-green lines, broken in a dot and dash manner or sometimes continuous lines, lying between the parallel to the clear veins. There is often one or more wide, dark-green streaks running down the outside of the leaf stalk near its junction with the pseudostem.

Control

1. Care must be taken to plant disease-free suckers.
2. Diseased plants must be destroyed as soon as the disease is detected.
3. To kill the aphid, kerosene oil should be poured to penetrate the funnel leaf and trickle between the leaf bases and the pseudostem right down to ground level. The whole stool including infected plant should then be dug out and chopped into small pieces to facilitate drying.
4. A solution of 1% 2, 4-D a weed killer should be injected at a rate of 15 ml. per plant at a height of 30 cm above the ground level.
5. Injecting the herbicide Agroxone 1(2, 4-dichloro-phenoxy acetic acid) or MCPA (2-methyl 1-4 Chloro-phenoxy acetic acid) into the diseased plant has been found effective in eradicating such plants in Australia (Rangaswami, 1984).

4.3.2 Banana Mosaic

This disease is prevalent in some parts of Maharashtra and was first reported in 1949 by Capoor (Rangaswami, 1984).

The disease is characterized by typical mosaic symptoms on the leaves. Diseased plants are stunted, the leaves reduced in size and clustered. Rarely a diseased plant produces fruits.

The disease is transmitted by the cotton aphid, *Aphis gossypii* Glover but not by the banana aphid, *Pentalonia nigronervosa* Cog.

The virus is of non-persistent type and it has practically no incubation periodin its vector. It is distinct from the virus causing the bunchy top, and is localized in some parts of Maharashtra, Gujarat and Karnataka States.

4.3.3. Heart rot or Infectious Chlorosis

The most general symptom is a chlorotic condition of the younger leaves formed by light green to yellow streaks of bands which extend out from the midrib. These streaks may be narrow and interrupted so that a mosaic effect is produced. During the colder months a soft, black rot may involve the funnel leaf and extend down into the heart of the plant. If this rot reaches the corm the whole plant may die.

The virus has other hosts like cucumber and tomato.

Control

When a symptom is observed, kerosene or an insecticide should be applied, and as soon as this has completed its work the whole stool should be dug out and chopped off.

DISEASES CAUSED BY FUNGI

4.4. PANAMA DISEASE

Panama disease or banana wilt, which occurs widely in Central America and West Indies, has largely been responsible for abandoning banana cultivation in thousands of acres. The disease was first noticed in Puerto Rico in 1903 and the causal organism *Fusarium oxysporum* f. *cubense* was described in 1910 (Tandon, 1967).

In India, the *Fusarium* wilt is an important destructive fungal disease spread in certain areas of Tamilnadu, Kerala and Karnataka states. The Indian strains of the pathogens could survive water stagnation for a month. However, longer periods of inundation (6 months to 2 years) have been found to completely eliminate the infection and restore the infested soil, making it fit again for successful banana cultivation.

4.4.1 Panama Disease

Isolation and preliminary characterization of *Fusarium oxysporum* from banana (Robusta) variety was done by Ajmeri *et al.* (2001). They also showed the disease development.

The presence of disease is indicated by the development of a deep yellow colour round the margin of the lower leaves which later turn brown and dry out. The leaf stalk collapses, leaving the dead leaves in a gradually increasing number draped round the pseudostem.

A definite diagnosis of Panana disease is made by splitting up the base of the plant lengthwise, when the corm will be found discoloured by numerous brown to black lines running in all directions through the white tissue. The brown vessels can often be followed up through the sheathing leaf bases and out into the vertical partitions of the leaf stalk.

The fungus is reported to consist of several strains which vary in their cultural characters and virulence. Ramakrishnan and Damodaran (1956) have reported existence of three different strains of the pathogen in Tamilnadu, differing only in their cultural characters and not in their pathogenicity.

Control

To control wilt is both difficult and costly,

1. Eradication of the stools has given partial control.
2. Chemical methods are expensive.
3. Rotations also as measure of control is not practicable as the funguspersists for long periods in soil.
4. Ramakrishnan and Damodaran (1959) reported that timing of pot soilreduced the viability of the pathogen to two months while in the control the viability extended up to four months.
5. Several fungicides are applied to eradicate *Fusarium* from soil. All the fungicides except Vapam, Thiram, and Mercuric chloride were found to be ineffective. Vapam at 8 oz in 8 gallons of water was most effective. Mercuric chloride at 2000 to 3000 ppm also seemed to control the disease (Sohi, 1983).
6. The most effective measure of control is cultivation of resistant varieties. At the Central Banana Research Station Aduthurai, Tamil Nadu, commercial varieties are being hybridized with the resistant wild species like *M. dalbisiana, M. acuminata* and *M. coccinea*. As an immediate measure of control certain resistant commercial varieties can be introduced profitably into the affected areas.
7. Care in cultivation, avoiding root injury and control of nematodes would also prove helpful in the control of the disease.

4.5. PSEUDOSTEM ROT OR HEART ROT

Pseudostem rot is caused by species of *Botryodiplodia, Gloeosporium* and *Fusarium*. Observations show that Poovan (Champa) variety is very succeptible to Pseudostem rot.

In the diseased plants, the inner leaves rot and die, and thus interfere with the emergence of new leaves. The rot gradually travels down towards the rhizome and prevents the emergence of inflorescence.

Control

1. Plantation sanitation, good drainage and adequate spacing reduce the incidence of the disease.
2. Timely spraying the plants with Bordeaux mixture is also helpful for preventing the spread of the disease.

4.6. LEAF SPOT OR SIGATOKA DISEASE

The disease is common in India, West Indies and Fiji. In India it is common in Tamilnadu, Kerala, Bihar and Bengal states and attacks mostly the leaves. Roy and Sharma (1952) reported it from Bihar where it was responsible for the death of plants.

It is distributed throughout the banana-growing regions of the world, with the exception of Egypt, Israel and Canary islands (Feakin, 1972).

After death of plants

Thammaiah *et. al* (1999) studied occurrence of Sigatoka leaf spot in high density population of Rajapuri banana in South India.

The symptoms caused by *Cercospora musae* Zimm. include appearance of bellow or pale yellow streaks, clearly visible on both the leaf surfaces (more prominent on upper surface). They consist of narrow, oblong, or elliptical areas, brown to black in colour, and about half an inch long by an eighth inch width. With age the centre dries out, leaving a characteristic grey spot bordered with a black line and surrounded by a narrow yellow halo. If the spots are numerous the individual lesions coalesce to form large peninsula of dead tissue extending from the margin towards the mid rib. Eventually the whole leaf dies. Minute grey tufts of fungal spores can sometimes be seen on the surface of the spots following prolonged rainy wether. In severe cases of infection, no bunch is formed or if there is any, the fruits are useless. During rainy season the spread of the disease is very rapid. Weak plants are easily infected and badly attacked. Sigatoka is common on poor badly-drained soils and in sheltered areas since in such a situation the humidity within the plantation is increased. Cultivation practices which increase plantation humidity, *i.e.* close planting, heavy weed or grass cover and failure to remove suckers, favour the built-up of the disease.

Control

1. Removal and destruction of the infected leaves are recommended forthe control of the disease.
2. Bordeaux mixture and some other oil based copper fungicides have been found effective against the disease.
3. Application of Copper oxychloride, Malachite green, Zineb and cuprousoxide at 3 weeks interval starting from December can reduce the damage.

4.7. LEAF SPECKLE

The speckle is caused by a fungus (*Mycosphaerella musae*). It is found on the under surface of the leaf as scattered or aggregated dark blotches, brown to black in colour, and of varying size and intensity. These dark patches are formed in the first place by a close speckling of the surface with grey dots, which later darken and coalesce. Unlike leaf spot the development of speckle is favoured by a shaded.

Situation and by close planting. The portion of lowest leaf nearest to the pseudostem is the first to the affected.

Control

Ground spraying of chemical suggested in Leaf spot or Sigatoka disease (4.6) to ensure a sufficient coverage on the lower leaves.

4.8. CORDANA OR LARGER LEAF SPOT

The disease is most prevalent in North Queensland, where it often causes extensive leaf decay in young plantations (in the first year) growing in sheltereds sites on scrub soils. The disease occurs also on the native banana (*Musa banksii*) which can be a source of infection to new plantation.

The spots are caused by a fungus *Cordana musae* (Zimm.) von Hohn. As the name suggests, the spots are more conspicuous than those of common leaf spot disease. When mature, they appear as large, elliptical or diamond shaped areas coloured with zonate shading on the upper surface and greyish on lower. They are bordered by a conspicuous band of bright yellow merging into green. The spores of the fungus, by means of which the disease is spread, are produced on the under surface of the spots where the fructification form a smoky : grey coat or concentric, black bands.

Control

By application of Bordeaux mixture and fungicide as suggested earlier.

4.9. HELMISTHOSPORIUM LEAF SPOT

The spotting caused by *Helminthosporium gibberosporium* Cruzi starts as deep seated necrotic lesions on the leaf veins. These are distributed throughout the leaf blade, but are more common towards the mid-rib. The necrotic spots soon develop into linear spindle shaped spots with pointed ends, few cms. in width and several cms. in length. The centres of the spot are grey to black, but not uniform in colour, and the thin margin is brownish to black, spreading along the veins. The spots ae generally surrounded by an orange-yellow discolouration, about 1 to 2 mm in width. Occasionally, several spots may coalesce to form large dried up patches.

Control

The disease can be controlled by two or three sprays of a copper fungicide.

4.10. BLACK TIP

The fungus *Helminthosporium torulosum* (Syd.) Ashby produces leaf spots and black tip of fruits. The damage is more severe when the fruit is involved.

Elongated spots with grey centres are produced on the leaf. The spots enlarge to form large patches, blighting portions of the leaf blade. It infects the fruits mostly at the tip, starting from the floral remnants. The infected tissue turns black and the infection is confined to the sk in, the pulp seldom being affected. Infection of the fruit causes uneven ripening.

Control

1. The diseased fruits in a bunch must be removed.
2. Preventive sprays with fungicides yield good result.

4.11. PESTALOTIOPSIS LEAF SPOT

Sohi and Prakash (1978) reported a new spot disease of banana caused by *Pestalotiopsis chethallensis* Sohi and Prakash. Earlier *P. laprogena* Speg. has been reported on banana causing fruit rot and spot (Merdith, 1963). Other species on *Musa* sp. include *P. funerea* Desm. on plantation leaves in Puerto Rico (Wardlaw, 1972), *P. fusescens* var. *sacelari* Waller on the surface of Gros Michel leaves in Jamaica (Ashby, 1971). *P. versicolor* (Speg.) Steyaert has been reported on Cavendish banana variety robusta causing leaf spot disease from our country (Prakash and Singh, 1975).

The spots were characterized by the presence of ashy coloured, circula-irregular areas originating from the tips or the margins and gradually advancing towards the base, becoming dark green coloured necrotic zone 0.5-15mm in length and upto 6 mm in breadth. Later on, they coalesce and from elongate or irregular big spots which can easily be seen from a distance. Severe infection were caused by splitting of the leaves. Healthy and diseased portion were demarcated by deep reddish yellow bands. Small pustules (acervuli) were invariably present on the upper surface of the lamina. Older leaves were more susceptible to infection as compared to younger ones. Consequent upon infection the whole leaf turned brown and started drying.

The species is named after the place Chethall from where it was discovered.

4.12. ANTHRACNOSE

The disease is also called summer black end and is caused by *Gloeosporium musarum* Cooke et Massee. The disease is serious in all the banana growing areas, especially in Bihar (Thakur and Misra, 1964; 1966).

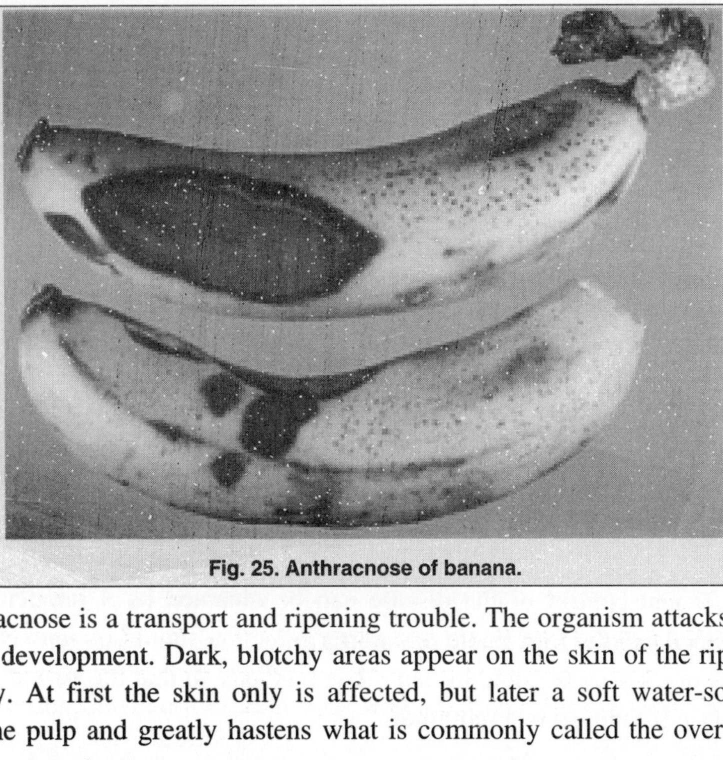

Fig. 25. Anthracnose of banana.

The Anthracnose is a transport and ripening trouble. The organism attacks the plant at all stages of their development. Dark, blotchy areas appear on the skin of the ripening fruit and enlarge rapidly. At first the skin only is affected, but later a soft water-soaked condition extends into the pulp and greatly hastens what is commonly called the over-ripe condition.

Black end or fruit stalk rot also affects the ripening fruit and occurs principally when the hands are broken into singles for packing. As the fruit ripens a soft, black and usually wet rot commences at the broken end of the fruit stalk.

All the varieties of banana have been found to be susceptible to the disease but its intensity varies. Roy and Sharma (1952) reported the differential reaction of the disease on different varieties. All the table varieties particularly Harichhal and Malbhog were severely attacked and at certain places banana cultivation had to be abandoned altogether due to the severity of the disease.

Misra and Singh (1962) studied the effect of temperature and humidity on disease development in Bihar. Optimum germination of conidia was obtained at 30°C. Maximum number of conidia germinated within 4 hours at this temperature and the germ tubes also showed a faster rate of growth. Optimum temperature for disease development was 30-53°C. The conidia germinated within a range of 85.7-100 per cent relative humidity. In Bihar the disease was most common during the months of June to September, when the temperature was high and was accompanied by frequent showers. The disease seemed to be arrested during cold as well as hotand dry months (Thakur anf Mishra, 1964).

Conrol

Spraying with Bordeaux mixture, Dithane Z-78 or Difolaton (0.2%) gives good control.

4.13. QUIRTER AND WINTER BLACK END

Affected fruits are a total loss and the difficulty of detecting the disease from a superficial examination of the case tends to make quirter the most serious transport disease of Australian bananas. The disease is caused by *Nigrospora sphaerica* (Sacc.) Mason which exists in non-parasitic manner on the surface of drying leaf-bases, the bunch spathe and other dead banana material in the plantation and on discarded bunch stalks and rotting fruit in the dumps near the packing shed. Tandon (1967) reported *N. oryzae* (Berk. & Br.) Petch. from the leaves from Savada and Hajipur and from China fruit from Hajipur.

The affected tissue has a dark, water-soaked appearance especially bordering the centre. The infected fruit has pulp decomposed to a dark semi-fluid state so that a squeeze of the hand will expel it in a stream from the stalk end or from the side. After the fruits have been broken from the stalk the fungus gains an entrance through the broken end and travelling up the fibres, sets up the typical rot in the pulp of the fruit.

Control

A very efficient control of the disease can be obtained by a fungicide Salicylanilide. During the winter months fruit liable to squirter infection should be marketed without delay and ripened as quickly as possible by upto-date methods so that the period during which the rot can take place is reduced to a minimum.

4.14. CIGAR-END DISEASE

Dade (1926) reported Cigar-end disease of Cavendish bananas due to *Stachylidium theobromae.* Holmes (1930) and Wardlaw (1931) also observed this disease and reported *S. theobromae* to be the causal organism. Shephard (1931), however, reported that Cigar end disease of Cavendish bananas in Mauritius was due to *Verticillium* species, particularly *V. theobromae*. Srivastava *et al.* (1964a) collected Cigar end of Mauritius bananas from banana orchards of Hajipur Bihar in the month of March 1963. The organism, associated was identified as *Gloeosporum musarum* Cooke et Massee.

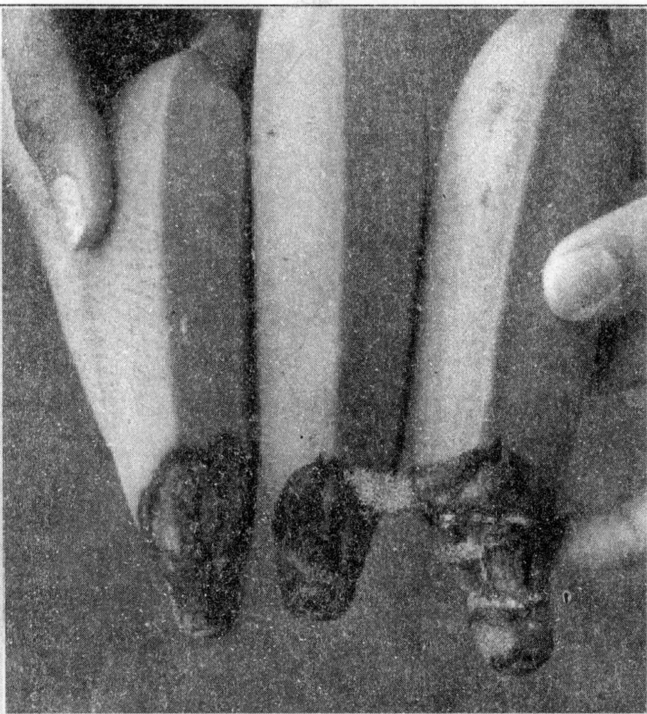

Fig. 26. Cigar end disease of banana.

Cigar end is a trouble affecting relatively young fruits in the plantation. Infection generally starts from the tips of the finger (stylar end) and slowly spreads backwards all around the finger causing greyishness of the skin. Shrinkage of tissues is consistent in severe cases. The diseased area later becomes covered with salmon to brown coloured fruiting bodies of *G. musarum.* The diseased surface appears corrugated showing close resemblance to burnt cigar tip. There is a sharp line of demarcation between healthy and diseased tissue. Usually no further extension takes place, but the fruit ripens prematurely. Usually no further extension takes place, but the fruit ripens prematurely. Usually only a few fingers show the disease, but occasionally a large proportion of the fruit on a bunch is affected.

Srivastava and Tandon (1971), while discussing postharvest diseases of banana in India, reported association of *Verticillium theobromae* (Turc.) Mason et Hughes with cigar- end rot from Bhusaval and Hyderabad during 1964-65.

Occasionally *V. theobromae* is associated with *Deightoniella torulosa* (Syd.) Ellis and produces tip rot of banana which bears some resemblance to cigar-end rot. In such cases generally one side of the finger is often more affected than another. The proximal end of the affected area is fawn- coloured dry and fibrous, whereas the distal end is dark coloured and a little crumbled. *D. torulosa* is associated with less decayed peel tissue and *V. theobromae* with extensively dry rotted pulp.

Both the organisms could infect bananas artifically but they failed to produce cigar-end symptoms either singly or in combination.

4.15. BLACK FINGER

Black finger is caused by a fungus *Dothiorella gregaria*, which forms small raised pustules on the skin. These exclude a minute tendril of spores producing a white, speckled appearance.

Like Cigar-end, black finger also is a disease of the young fruit. While the fruit is still immature and angular one or more fingers may develop a jet black decay commencing at the tip and extending back towards the base until, unlike as in the case of cigar-end, the whole fruit is involved. The fruit becomes tapered by the gradual shrinkage of the affected region, which remains firm and eventually dries up to form a mummy.

4.16. DEIGHTONIELLA FRUIT SPOT

During March, 1962 some Banshibat variety of bananas were collected from banana orchards of Hajipur (Bihar) (Srivastava, 1964a). *Deightoniella torulosa* (Syd.) Ell. (Syn. *Helminthosporium torulosum* Syd.) was isolated from the rotten fruits. This pathogen is also associated with Cigar-end rot (4.13).

Infection appears in the middle or from the tip portion of the fruits and extends longitudinally both at the proximal and distal ends of the fingers. This portion gradually increases and covers the entire surface which is exposed to sun. Conidiophore and conidia of *Deightoniella* are readily isolated from these portions. Occasionally in those cases where the fruits become withered off a species of *Fusarium* is also associated. There is a marked distinction from the green healthy tissue with slightly sunken grey of black diseased area.

4.17. LASIODIPLODIA ROT

The disease is caused by *Lasiodiplodia/Botryodiplodia theobromae* Pat. Tandon (1967) isolated it from the fruits collected in Patna (Bihar). It also occutrred in the leaves of Chini champa from Jorhat and Kanthali from Chennai and Bhusaval. The disease was found from Allahabad, Kanpur, Delhi, Bhusaval, Savada (M.S.) and Hyderabad (A.P.) by Srivastava and Tandon (1971).

At an early stage the diseased fruits show a brownish discoloration of the rind. As the lesion increases in size it becomes somewhat irregular in outline and assumes a water soaked appearance. Often the infection starts the distal end of fruit and the rotting gradually progress upward to the stalk end. The whole fruit becomes dark brown in colour. The infected region collapses and shrinks and numerous dark coloured pycnidial bodies of *B. theobromae* develop all over. Williamson and Tandon (1966) studied the effect of temperature and humidity on the development of disease. Maximum rot was produced at 25°C. External supply of humidity enhanced the spread of rot. Percentage germination of spore was more at 25°C. Of the various nutrient solutions tried, in 0.5% and 1% glucose solution germination of spores was 91% after 6 hours. Cell wall degrading enzymes of *B. theobromae* were studied by Gupta and Vyas (1979) and they found that the activities of both Polygalacturonase and cellulase enzymes produced by less virulent isolate were much more remarkable as compared to those of highly virulent isolate. Pathogenic potentialities of the two isolates are not directly related with their ability to produce cell wall degrading enzymes.

Control

Bhargava *et al.* (1983) tried two new fungicides, Delan and Saprol, on Bombay Green variety and found Delan (dithinanon 2,3-dicarbonitrile-1, 4-dithiaanthraquinone) more effective.

4.18. BLACK PIT

The trouble is caused by a fungus *Piricularia grisea* (Cook) Sacc. The disease is sporadic in occurrence. Black pit makes its appearace during the wet part of the summer and generally persists through the cooler months.

The disease commences as small reddish spots on the green skin of fruit approaching maturity. These spots develop into typical shallow black pits. The spotting is most abundant on the upper hands of the bunch and has been observed on the bunch stalk and on the soft succulent sheath-like stalk of the transition leaf. Although the lesions are restricted to skin yet the fruits with numerous pits are ugly to look at and at times whole bunches have to be discarded.

Control

Spray of copper oxychloride reduces the incidence of disease.

4.19. CHAETOSPHAEROPSIS ROT

Rai and Lal (1984) reported occurrence of a new fruit rot in the markets of Allahabad during July-December, 1976. The pathogen responsible for the disease is named as *Chaeotosphaeropsis truncata* var. *indica* (Rai and Lal).

Generally the infection occurred at the stylar end of the fruit as a brownish and water-soaked discolouration of the peel spreading around the point of incidence and the rotting progressed upwards eclipsing the entire fruit within 8-10 days. Then colour of the skin turned black. In severe infection the skin shrivelled and numerous dark coloured pycnidia of the pathogen appeared all over the surface of the fruit. It was also observed that the rotting of the pulp was faster than the peel.

Control

Srivastava *et al.* (1988) tried four fungicides against the pathogen *in vitro*. They found Delan uneffective against the pathogen. All the rest fungicides were tried in three concentrations *in vivo*. Calixin at 1000 ppm was most effective (90% disease control) followed by Tecto-40 (80%) and Dithance M-45 (30%). Earlier Rai (1982) working with the same organism has recommended the use of Benlate for controlling the rot of banana.

4.20. PINK MOULD ROT

It is caused by *Trichothecium roseum* Link ex Fr. (Srivastava and Tandon, 1971).

The disease is characterised by the production of pink mouldy growth over the surface of the fruit. The rot never extend to the pulp and is only limited to the peel of the fruit. Inoculation with *T. roseum* revealed that it is a weak parasite and could infect the fruit only if they were previously injured.

4.21. BLACK HEART OF BANANA FRUITS

The disease is caused by *Fusarium moniliforme* Sheldon and *Fusarium roseum* Link. (Rao, 1966; Peshney *et al.*, 1984).

During September-October when bananas were midway towards maturity, few fingers from the bunches even though normal in colour appeared to be reduced in size starting from pistil end. Such fingers when cut open showed dark brown discolouration extending from placenta to the pulp and from pistil end to stalk end. Ultimately the whole tissue turned brown and necrotic skin was green and intact.

Control

1. Removal of pistils followed by spraying with Carboxin, Oxycarboxin(each 0.1%) and Ziram (0.25%) were found effective (Pashney *et al.*, 1984)
2. Benomyl and Aretan were toxic *in vitro* to *F. moniliforme* and *F. roseum.*100 ppm Benomyl successfully checked rots caused by the two pathogens.
3. Aureofungin was found effective only at higher concentration (2000ppm by Khanna and Chandra (1976).

4.22. RHIZOPUS ROT

Tandon and Mishra (1969) found *Rhizopus stolonifer* (Ehrenb. ex Fr.) Lind. causing fruit rots of *Carica papaya* and *Musa paradisiaca.*

The disease causes softening of tissue. Outer skin remains intact in initial stages but when completely rotten the whole tissue turns soft and pulpy.

Control

The disease can be controlled by storing the fruits at 10°C. Isothan Q-15 and Nickel containing fungicides were effective *in vitro* but failed to control the disease *in vivo* (Tandon and Mishra, 1969).

4.23. CERATOCYSTIS ROT

Ceratocystis paradoxa is another major cause of banana wastage (Andrade *et al.* 1956), which is called black head disease or cushion rot or crown rot. The rot exhibits symptoms like finger dropping, blackened and softened tissue with characteristic sweetish odour. The fungus grows out of epidermis and forms cob-web like mass of dark coloured mycelium. It also induces premature ripening and rotting of main stalk in the proximity of hand while the fruits are still unripe (Wardlaw, 1972).

4.24. SEPTORIA LEAF SPOT

It is a new leaf spot disease caused by *Mycosphaerella enmusae* (anamorph *Septoria enmusae*). It is found on under surface of leaf (Caulier *et al.* 2000)

4.25. TABLE. SOME MORE PATHOGENS ISOLATED FROM MUSA PARADISIACA L.

Sr. No.	Organism	Substratum	Reference
1.	*Alternaria tenuis Auct.*	Leaf	Tandon, 1967
2.	*Phyllosticta sp.*	Leaf	"
3.	*Macrophomina phaseoli (Maubl.) Ashby*	Leaf	Tandon *et al., 1965*
4.	*Melanospora damnosa (Sacc.) Lind*	Fruits & Leaves	Srivastava *et. al., 1964a*
5.	*Phoma sp.*	Leaf	"
6.	*Curvularia lunata (Wakker) Boedijn*	Leaf	Tandon, 1967
7.	*Phyllosticta gastonis Roum.*	Leaf	"
8.	*Cephalosporium sp.*	Leaf	"
9.	*Memononiella echinata (Riv.) Gall*	Rhizome	"
10.	*Helminthosporium sp.*	Penducle	"
11.	*Cochliobolus spicifer Nelson*	"	Tandon, 1965
12.	*Pestalotia leprolegna Speg.*	Leaf Base	Tandon 1967
13.	*Fusarium sp.*	Flower	"
14.	*Fusarium sp.*	Spath	"
15.	*Fusarium sp.*	Fruit	"
16.	*Curvularia falax Boedijn*	"	"
17.	*Malustela aeria Batista et al.*	"	"
18.	*Hendersonula toruloidea Nattrass*	"	"
19.	*Rhizopus nigricans Ehreb.*	"	"
20.	*Papularia sphaerosperma (Pers.) Hohnel.*	"	"
21.	*Curvularia lunata (Wakker) Boedijn*	"	Ved Ram & Dharma Vir, 1986.

4.26. BACTERIAL DISEASES

4.26.1. Moko Disease

Moko disease, a bacterial wilt disease, was first recorded by Schomburgh in 1840 in Guyana and the first description was given by Rover (1911 c.f. Rangaswami 1984). The disease is caused by *Pseudomonas solanacearum*, E. F. Sm. The disease has been recorded in Venezuela, Trinidad, Guyana, Costa Rica, Honduras and Panama. In India Chattopadhyaya and Mukhopadhyay (1968), and Ramgaswami (1984) reported Moko disease in West Bengal. The disease is suspected to be prevalent in Kerala and in North Arcot district of Tamil Nadu.

The infection occurs through the above ground parts, and mainly through incidental wounds. In the initial stages the bacterial wilt is characterized by the development of a yellowish discolouration of the inner leaf lamina close to the petiole. This is followed by the collapse of the lamina near the junction of lamina with petiole. Within a week, most of the leaves show wilt symptoms, and consequently, wilting of the inner leaves with the exception of heart leaf.

When a diseased pseudostem is cut transversely it is found that all vascular strands are discoloured, the colour ranging from pale yellow to dark brown or bluish-black, clearly indicating its vascular nature. The reddish tinge present in typical cases of Panama wilt is usually absent. The infected tissues exude dirty white, grey or brown slimy bacterial drops. The Moko infected pseudo-stem has no smell when cut, unlike those of Fusarial wilt.

Control

1. All the tools used for pruning and cutting should be disinfected with formaldehyde (one part of formalin in three parts of water).
2. Since the insects can carry Moko bacterium to the male flowers, the removal of male flowers as soon as the last female hands are emerged on varieties with dehiscent bracts, would help in minimising the spread.

4.26.2. Bacterial Leaf Blight

A bacterial disease of banana leaves caused by *Xanthomonas musarum* Rangaswami and Rangaajan has been reported to occur in a severe form in Tamil Nadu (Rangaswami, 1984).

The diseasee is characterized by black streaks along the veins. The severely infected leaf may roll up, and in some cases the infected portions give way, resulting in distortion of the blade. The leaves dry up and wither and the diseased plant becomes distinct, even from a distance.

5

Ber

5.1. GENERAL ACCOUNT

The Chinese date, Chinese fig, ber or bor is an important minor fruit of India. It is the most hardy fruit-tree cultivated all over India and is often called the poor man's fruit. *Z. mauritiana* is ever green, and is commercially most important fruit in India, where it is known as Indian jujube or ber.

Most of the trees grow isolated, are of poor quality and their fruit sells cheap. However, many plantations of budded trees have now come up all over India, and fetch higher prices than even grafted mangoes. The ber (*Zizyphus mauritiana* Lamk.) due to its xerophytic character, can be grown economically on marginal soils and can withstand high alkalinity. It occupies about 12,000 acres, distributed mostly in Bihar, Gujarat, Assam, Andhra Pradesh, Tamil Nadu, Punjab, Uttar-Pradesh, Madhya Pradesh and West Bengal.

Jujube fruits have a spongy, sweet-tasting pulp, and are an excellent source of ascorbic acid and carotenoids (Bal and Mann, 1978). They are generally eaten fresh. In Iraq they are more expensive than most other fresh fruits. The fruits are drupe characterized by a highly lignified and pointed endocarp (stone or pit), fleshy mesocarp (pulp) and a thin pericarp (skin). There are, however, some cultivars such as **Mallacy,** in which the endocarp is thin.

The Indian varieties are not as rich in sugar and Vitamin C as the good Chinese varieties. However, they contain 50 to 150 mg of ascrobic acid per 100 g of fruit and 12 to 18.7 per cent sugar. This is even better than the Vitamin C content of the juice of sweet oranges. The candied ber also makes an excellent product. Ber is also used for rearing the insect *Tachardia lacca*, or *Laccifer lacca* Kerr, which yields lac (Shellac).

Ber belongs to the genus *Zizyphus* of the family Rhamnaceae, which is very close to the family of grapes, Vitaceae. Several spp. grow wild in India like *Ziziphus oenoplia, Z. rugosa, Z. xylocarnus* and *Z. rotundifolia*. The cultivated Indian ber is very distinct from the Chinese ber, which belongs to the species of *Z. jujuba*. The Indian ber has a spreading tree, vine like branches, leaves which are dark green on the upper surface and densely felted on the lower surface. It flowers in autumn, bears fruits at the end of winter, if it sheds the leaves it does so in the hot weather after fruiting and does not like a cold climate. These characteristics are not found in the Chinese species. The Indian species is *Z. mauritiana* (Ranjit Singh, 1985).

Varietal selections carried out so far has given some promising varieties for commercial growing. In the Punjab, Umran, thornless, Kalagola have been found promising, though Kaithli, Dandan and Chonchal are also grown to a certain extent. In Uttar Pradesh, Banarsi and Gola are favoured most, whereas, in Bihar the important varieties are Banarsi, Nagpuri and thornless. In Gujarat and Maharashtra, seedless Bor is preferred over others. Gola of Delhi is the earliest to come into Delhi market and fetches a high premium.

Jujube fruit will ripen either on trees or after harvest, provided the picking is done at the proper stage of maturity. Jujube fruits ripened on the tree generally have a short storage life, and the best results are obtained if they are picked before the onset of ripening. Jujube fruit have a short life (6-8 days depending upon cultivar). Experience and researches have shown that the specific gravity of the fruit and fruit colour are more suitable maturity indices. According to Mathur (1996), Ber cultivation has started in Rajasthan after water of Indira Gandhi canal project has reached. A six-year old of Umran yields 80-120 kg ber giving a net profit of Rs. 32,000 to 36,000/ha.

5.2. BIRDS AND INSECT PESTS

Birds attack the fruits of ber as much as they attack the guavas. During the fruiting season the ber trees are covered with fishing nets in some places to protect them from birds. Some of the common insect pests on ber are as follows :

5.2.1. The Ber Fruit-fly : *Carpomyia vesuviana Costa*

(Family-Tephritidae, Order-Diptera)

The insect is prevalent in India, Pakistan and the Middle East extending upto Southern Europe. It is the most important pest of ber in North-West India. This is one of the important pests of ber fruits (*Zizyphus*) as nearly 80 per cent of fruits are rendered unfit for consumption. It is active early in spring and again in autumn, and the summer is passed in pupal stage. Its incidence is serious during February-March. The fly oviposits on semi-ripe fruits. The maggots feed on the pulp and pupate in the soil underneath the tree.

The deformed and spoiled fruits with minute holes and dropped ones should be destroyed. Clear-off the wild ber bushes around the orchards. Dig the basins of the tree with spades during summer to expose the dipausing pupae to the bright sun and birds. Spraying of Dimethoate (0.05 per cent) or Carbaryl (0.2%) twice or thrice at monthly intervals commencing from the pea-sized fruit stage is an effective control measure (Reddy *et al.*, 1996).

Control

In Madhya Pradesh a chemical control schedule of three or four spraying at triweekly interval commencing from last week of November is done. The first1 two with lindane 0.1 per cent (before ripening), the third with m alathion 0.06 per cent (during ripening) and depending on necessity a fourth with malathion 0.06 per cent has been recommended (Nayar, *et al.*, 1981.

5.2.2. Tree Hopper or Cow Bug : *Tricentrus bicolor Dist*

(Family-Membracidae, Order-Hemiptera)

This is widely distributed in India. It is characterized by two Ocelli. Antennae lie in a depression beneath and in front of the eyes. They measure 10 to 12 mm. stridulatory organs absent. They usually jump. They are provided with pronotum extending backward over abdomen and ornamented with various spines. Wings are largely concealed by pronotum. The apex is prolonged posteriorly into a robust posterior spine.

They suck the sap from trees. Principal damage is caused by egg laying. Soon after copulation the female selects the lower surface of the tender branches, not exposed to the sun directly and oviposits. Sometimes egg may be laid along the midrib of the leaf, petiole, buds or even leaf lamina. The female cuts a longitudinal slit on the bark and inserts the eggs, which may be laid singly or in groups.

Control

Lindane, Phosphamidon and malathione are effective.

5.2.3. Lac Insect : *Laccifer lacca*

(Family-Lacciferidae : Order-Hemiptera)

The Indian Lac insect *L. lacca* has a wide distribution in India, and the resinous secretion of the females provides the stick-lac, which is of great commercial importance. The apterous females are highly degenerate with vestigial antennae. The body is irregularly globose and enclosed in a resinous mass secreted by the lac-resin glands distributed all over the dermis except near the two branchial pores and the anal tubercular region. The encrustation covers the twigs of host trees to a thickness of 6 to 12 mm.

5.2.4. Leaf Webber : *Tonica zizyphi* M.

(Family-Occophoridae : Order-Lepidoptera)

These are small moths generally brownish in colour with apically rounded road wings. The Larvae generally web together or roll the leaves or sometimes feed on seeds. In India larvae of *Tonica zizyphi* roll or web the leaves of *Z. jujuba*, orange, lemon and *Murraya koenigii*.

Control

1. Cut and burn the infected leaves/twigs, or
2. Spray of DDT 0.2 per cent to kill the larvae.

5.2.5. Leaf Butterfly : *Tarucus theophrastus* F.

(Family-Lycaenidae : Order-Lepidoptera)

Small to moderate sized butterflies with upper surface of wings being metallic blue or coppery. Tail-like delicate prolongations are seen on hind-wings. Each eye is surrounded by a rim of white scales land the antenna ringed with white. Legs are normal excepting the fore logs of males which may possess more or less shortened tarsi, and may be wanting in one or both claws. The colouration in both sexes often may show great differences. Lycaenid larvae are characteristically onisciform with both ends tapering and the enlarged sides concealing the legs. They have numerous secondary setae on their body. Larva of *Tarucus theophrastus* feeds on leaves and buds of *Zizyphus jujuba*.

Control

Spray of 0.1 per cent DDT once in three weeks.

5.2.6. Fruit borer : Meridarchis scyrodes Meyr

(Family-Carposinidae : Order-Lepidoptera)

The well known pests of importance are *Meridarchis scyrodes* and *M. reprobata* with the larvae feeding on fruits of *Zizyphus jujuba* and *Syzygium jambolanum* respectively. It is reported from South India (Nayar *et al.*, 1981). Reddy *et al.*, (1996) reported it from rainfed areas of Andhra Pradesh. It is sporadic. Bred throughout the year the insects migrate from one variety to another. Besides, these are also found on jamun fruits.

Control

Spray of BHC reduces the disease/pest incidence.

5.2.7. Table : Some other Insect Pests of Ber

Sr. No.	Name	Family	Order
1.	*Leptocentrus obliquis W.*	Membracidae	Hemiptera
2.	*Machaerota planitae (Spittle bug)*	Cercopidae	Hemiptera
3.	*Nipaecoccus vestator Mask.*	Pseudoco-ccidae	Hemiptera
4.	*Lecanium discrepans Gr.*	Coccidae	Hemiptera
5.	*Pulvinaria burkilli Gr.*	Coccidae	Hemiptera
6.	*P. maxima Gr.*	Coccidae	Hemiptera
7.	*Aonidiella orientalis Newst.*	Diaspididae	Hemiptera
8.	*Chionespis megaloba Gr.*	Diaspididae	Hemiptera
9.	*Scirtothrips dorsalis Hood (Flower thrips)*	Thripidae	Thysanoptera
10.	*Dolichothrips indicus H. (Leaf thrips)*	Thripidae	Thysanoptera
11.	*Indrabela tetraonis M. (Bark-borer)*	Metarbelidae	Lepidoptera
12.	*I. quadrinotata W. (Bark borer)*	Metarbelidae	Lepidoptera
13.	*Synclera univocalis Walk. (Leaf roller)*	Pyrantidae	Lepidoptera

14.	*Taragama siva Lef.*	Lasicampidae	Lepidoptera
15.	*Thiacidas postica W. (Hairy Caterpillar)*	Noctuidae	Lepidoptera
16.	*Platypria andewsi W. (Leaf hispid)*	Hispidae	Coleoptera
17.	*Ceroplastodes cajani (Scale insect)*	Coccidae	Hemiptera

DISEASES CAUSED BY FUNGI

5.3. POWDERY MILDEW

This disease was first reported from Allahabad (Mitter and Tandon, 1930), Bombay (Uppal *et al.* 1935 ; Patel *et al.* 1949). Mehta (1950) reported it from Kanpur (U.P.). Kapur *et al.* (1975) reported occurrence of powdery mildew of ber in Punjab. All the 37 ber varieties tested were susceptible to disease (*Oidium* sp.) Recommended varieties Umran and Kaithi showed moderate to severe infection, while several others were tolerant.

The symptoms of the disease are noticed on flowers and fruit in 3rd week of October. The disease may appear earlier when conditions become favourable. The disease starts in the form of white powdery spots on the surface of fruits and later covers the whole area with the age. The spots change into light brown to dark brown discolouration. The infected area becomes slightly raised having canker and fruit crack later. The effect of temperature and relative humidity was studied on development of powdery mildew of ber (Maheshwari and Singh, 1999). Max. disease intensity was recorded in December. The lower disease intensity was recorded at high temp. and low RH.

This disease is caused by *Oidium erysiphoides* f. *zizyphi* Yen and Wang. The fungus survives in diseased debris which become the primary source of infection. The secondary infection takes place through spores (conidia are present in air).

Control

The disease can be controlled by spraying Dithane M-45 0.2%. Two or three sprays should be given after two weeks inteval. Of the 9 fungicides tested the best was benlate (92.1%) control followed by Afugan both at 0.07%. Suhag and Mehta (1983) reported participation of host in modifying the toxicity of some antipowdery mildew chemicals like Bavistin, Selfex, Karathane (Dinocap) etc. They found that leaf exudates modified the fungitoxicity of chemicals.

Oidium erysiphoides f.sp. *ziziphi* of ber *Z. mauritiana* cv. Umran can be controlled by triadimefon (Desai 1998 b). 10 different treatments were tried.). 1% tridimefon at flowering stage is recommended.

5.4. WILT OF BER

Wilt of ber seedlings is characterized by drooping of leaves then progressive death of plants. Two species of *Fusarium F. equiseti* is reported from Jodhpur (Lodha, 1984) and *F.*

oxysporum from Hisar (Gupta and Madan, 1985). The common cultivars like Umran, Gola and Kathali are susceptible (Chand *et al.*, 1986). The disease can be controlled by soil drenching with Bavistin and Calixin.

5.5. RHIZOPUS ROT

In case of soft rot caused by *Rhizopus oryzae* Went and Prisen Geerlings, light brown lesions appeared on the skin of the fruit which rapidly increased in size and the whole fruit got covered by white fluffy mycelial growth bearing black mature sporangial heads.

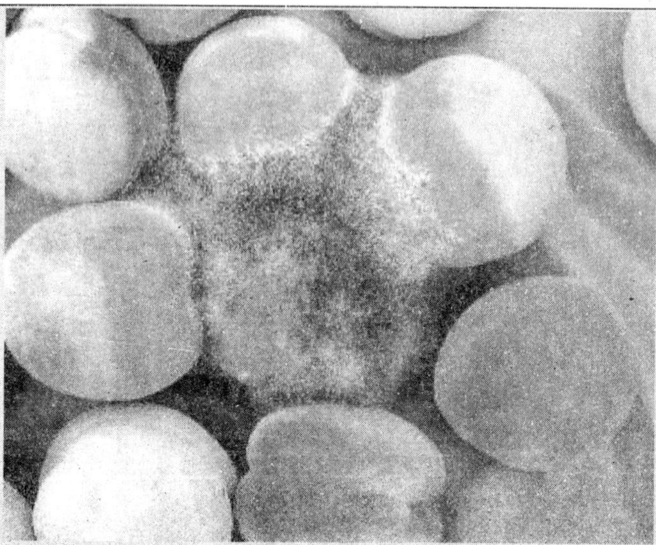

Fig. 27. Fruit rot of ber caused by Rhizopus oryzae.

5.6. PHOMOPSIS ROT

Lal *et al.* (1981) observed a serious soft rot of *Zizyphus mauritiana* Lamk. Var Pewandi caused by *Phomopsis natsfume* Hara in the local markets of Allahabad and Kanpur. Earlier the pathogen was found on leaves of *Z. nummularia* Wt. and Arn

Fig. 28. Phomosis rot.

The disease appeared as a light russet-vinaceous (brown) coloured irregular spot which increased further and finally the whole fruit turned pulpy, seal brown to black in colour with soft and loose outer skin. The mesocarp of the fruit also turned vinaceous brown as the disease progressed. In later stages the fruit was deformed and numerous pin head like pycnidia developed on the shrunken epicarp. The fruit completely rotted within 10 days. Injury was essential for the development of the disease and ripe fruits were found to be more susceptible than the green ones.

Of the five fungicides tried Difolatan at 1000 ppm and Bavistin at 500 ppm were most effective (90-95 per cent desease controlled). The other three fungicides Dithane M-45, Blitox and Benlate showed 60-85 per cent disease control. Pre-inoculation and post-inoculation treatment did not show any significant difference except in case of Blitox where pre-inoculation treatment was more effective.

5.7. PESTALOTIOPSIS ROT

During a survey of fruit rot diseases of ber in the months of January-March, 1977 in the local markets of Allahabad and its adjacent regions, a wide spread severe rot of ber was recorded by Rai *et al.* (1982), which was caused by *Pestalotiopsis palmarum* (Cooke) Steyaert.

Disease develops as a small rounded to oval water-soaked lesions which enlarge and become dark brown after 5-6 days. In the beginning white cottony mycelia appear on the diseased area which eventually covers the whole fruit producing black acervuli. Mesocarp also becomes black due to the production of acervuli and finally the infected fruit gets deformed. The stalk-end of fruit was more susceptible than the stylar-end.

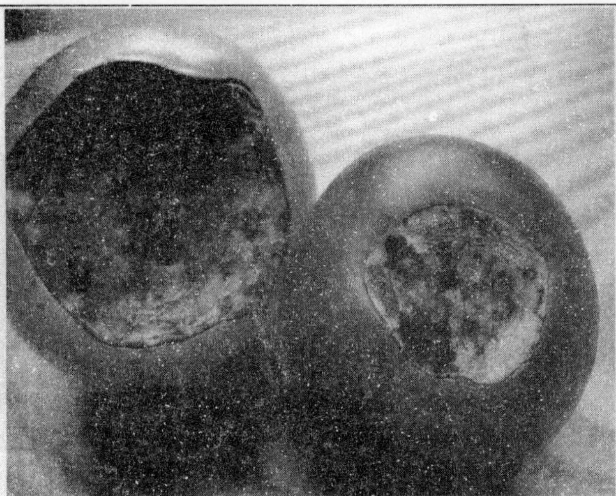

Fig. 29. Pestalotiopsis rot of ber.

Physiological studies on *P. palmarum* were carried out by Rai (1982). He found best growth at temp. 25°C, the growth was good between pH 4.0 and 7.0 and best at pH 5.5. Excellent sporulation was observed at pH 5.5 and 6.5. Best carbon sources were Maltose and

Cellobiose, best nitrogen source was L-glycine followed by L-histidine. Spore germination was 95% on host extract and 88% on 1% maltose. At 25°C 85% spores germinated. Spores germinated between pH 3.0 and 10.0, best germination was obtained at pH 5.5 (82% germination).

Control

Two fungicides Bavistin and Benlate and one antibiotic Aureofungin are effective against the pathogen.

5.8. LASIODIPLODIA ROT

Noticed during January-February, the disease was characterized by the production of irregular black lesions on fruits. In the beginning lesions were small, slightly depressed and dark brown in colour, subsequently they increased in size and cover the whole fruit which completely rotted after 10 days of infection and developed dark brown hyphae with small fruiting bodies all over the surface of the fruit.

Fig. 30. Lasiodiplodia rot.

The causal organism was identified as *Lasiodiplodia theobromae* Pat. (Rai *etal.*, 1982).

5.9. CERATOCYSTIS ROT

In case of Ceratocystis rot the decay was characterized by the appearance of small, light brown lesions which enlarged and covered the entire fruit. Ultimately the fruit became soft, dark brown and got covered by dark olive green and finally black crust of the fungus. Prominent black perithecia with straight and long necks were seen on the infected portion of the fruit. The pathogen has been identified as *Ceratocystis adiposa* (E. Butler) C. Moreau (Badyal and Sumbali, 1990).

5.10. CERATOCYSTIS ROT

Mukherjee *et al.* (1999) observed effect of pre-harvest application of some fungicides on the shelf life of ber. Storage quality of fruits of Umran variety improved.

5.11. BLUE MOLD

Caused by *Penicillium* sp. appeared as watery spots, shallow at first but laterextended deeper into the tissues. The surface of the spots become studded with tufts of mycelial growth which at first were white, but later turned bluish green due to the production of large number of spores.

Isolations were made and colonies of *Penicillium crustosum* Thom were obtained which produced white colonies on P.D.A. turning blue and finally grey (Badyal and Sumbali, 1990).

5.12. PHOMA ROT

The disease was observed in Horticultural orchard of Haryana Agricultural University, Hissar (Gupta and Madaan, 1977). The infected fruits were characterized by small, slightly depressed dark brown spots near the stem end. The erumpent pycnidia developed on these spots. Lesions were noticed on the ripe fruits. The lesions were irregular in shape. The disease is caused by *Phomu hissarensis* Gupta and Madaan. *Phoma macrostoma* was reported by Sidhu and Singh (1981). Another species *P. eupyrena* (Dang *et al.*, 1984) was reported on leaf and fruit.

Leaf spots are light brown coloured few mm in diameter becoming large oval or irregular. Black pin head like pycnidia develop after 15-20 days.

Control

Efficacy of fungitoxicants against leaf spot of ber caused by *P. eupyrena* was studied by Dang *et al.* (1986).

In pre-treatment Topisin-M gave complete control of disease at 0.1 and 0.2% concentration. In post-treatment also, it provided 89% at 0.1% and 100% at 0.2% followed by Bavistin and Blitox at 0.2%. Delan and Rovral were least effective. Efficiency of Bavistin against. *P. macrostoma* has been reported by Sidhu and Singh (1981).

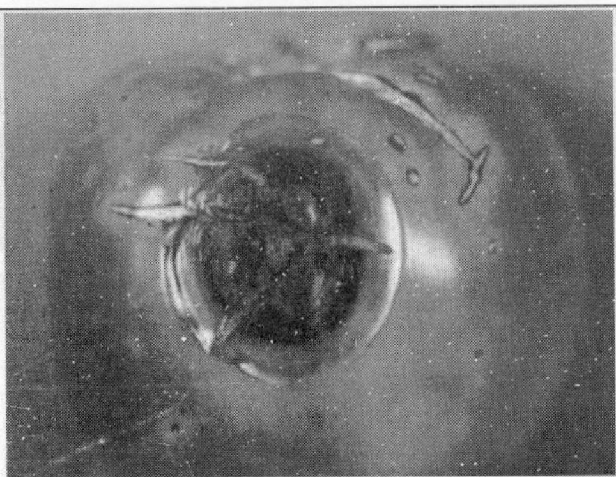

Fig. 31. Phoma rot of ber.

5.13. ALTERNARIA FRUIT ROT

The disease starts in the form of slightly depressed, brown to dark brown, circular lesions on the fruits. Sometimes concentric rings are also present on the spots. The smaller spots coalesce and form larger spots. The disease is incited by *Alternaria* state of *Pleospora infectoria* Fuckel.

5.14. COLLETOTRICHUM ROT

The disease starts with small, slightly depressed light brown, water soaked lesions on the fruits. With the age they enlarge and coalesce to form larger spots. Under humid conditions they produce acervuli on the spots in masses. This anthracnose disease is caused by *Colletotrichum gloeosporioides* Penz.

5.15. CLADOSPORIUM ROT

Bluish-green spots were visible on the ripe fruits in various stages. In most cases mechanical injury resulting in scratches or wounds was responsible for infection of the fruits.

Panwar & Vyas (1974) found this disease caused by *Cladosporium oxysporum* Berk. and Curt. This pathoen also infected fruits of *Punica granatum* L. and *Capsicum annum* L. in the local markets of Jodhpur City (Raj.). The pathogenicity of the fungus on above mentioned fruits was confirmed both by injury and contact methods. The mycelial growth was observed on the fruits 8-10 days after inoculation at 25-28°C. Cross inoculations were successful on the fruits of *Lycopersicon esculentum* Mill., *Citrus aurantifolia* Swingle, *Vitis vinifera* L. and *Pyrus* sp.

5.16. TRICHOTHECIUM ROT

This disease is incited by *Trichothecium roseum* Link.

The infected fruits showed pink colour spots on the fruit which covered a large area within a few days of infection.

5.17. EPICOCCUM ROT

During surveys (1999-2000) undertaken in arid region of Rajasthan a new fruit rot of *Z. mauritiana* var. Gola was noticed (Nallathambi *et al.* 2002). Infection on maturing fruits appeared as brown strips progressing downwards in one side of fruits, whereas other side was showing yellowing. In severe infection, depression of fruit surface, hardening and malformation of such fruits were observed. In some fruits with severe infection, small pin head like growth of the pathogen having clusters of dark conidia were found. Internal symptoms showed fibrous nature of pulp tissues with necrotic seeds. On isolation on PDA slants *Epicoccum nigrum* Link was isolated.

MYCOPLASMAL DISEASE

5.18. JUJUBE WITCHES BROOM

Witches broom like symptoms were reported on chinese jujube from China (Zhu *et al.* 1983). The leaves of *Z. jujuba* trees were found affected with mosaic like symptoms. Virus particles of 20 × 370 × 530 nm were isolated. Kim(1965) also reported similar observations from Korea. Pandey *et al.* (1976) reported witches broom like symptoms in *Z. mauritiana* seedlings only.

Pandey *et al.* (1976) from Pune suggested to control this malady throughtetracyclene at 500 and 1000 ppm concentration.

6

Sapota

6.1. GENERAL ACCOUNT

The sapota (*Achras sapota* L.), sapidilla, or bully tree, commonly called, 'Chiku' in India belongs to the family Sapotaceae (order-Ebenales of Gamopetalae). It is native to Tropical America. Like guava, it is so well established in India that the man in the street considers it to be indigenous to India. Smaller than apple theyhave a greyish brown skin and sweet, yellowish-brown, granular flesh. Sapodillas are usually eaten fresh, but are also delicious when baked with butter and sugar.

In India, it is grown only for its fruits, but in other countries the milky latex from the bark of the tree yield an important commercial product, gutta parcha, which forms the base for the manufacture of Chewing-gum.

In India, the main centres of its cultivation are the area between Mumbai city and Surat, the coastal areas of Andhra Pradesh around Vishakhapatnam, Chennai, Mysore, Saurashtra, the sub-montane area of Uttar Pradesh and Bengal.

In less known fruits like sapota, ber aonla and pomegranate area expansion and production rise during recent years is simply spectacular (Ghosh, 1999). India tops the world in sapota production and high productivity has been achieved in Karnataka (17.2 tonnes/ha), followed by Maharashtra (16 tonnes/ha).

It is not known when sapota was first introduced into India but from one record cultivation of this fruit species was taken up for the first time in Maharashtra in 1898 in a village named Golwad in Dhanu taluka of Thane district (Chadha, 1993). From another record a traveller from Mumbai Mr. Dinsha Petrit introduced two sapota plants from Singapore in the fag end of Nineteenth century and planted them near Govalia tank. From here Mr. Kawamji Patel and Mr. Ardesar Irani, two Parsi gentlemen introduced these to

Golwad village. It is from here that this fruit species spread to various parts of the country (Bari 1991).

Grafted chiku trees begin to bear fruit after three years. Since it is a slow growing tree, the yield varies from 1500 to 3000 fruits per adult tree. But in a variety like Cricket ball the number may be only 500 fruits (Singh, 1969).

Sapota fruit has good transport and keeping qualities. It is astringent till fully ripe, as in the case of some varieties of persimmon. Sapota fruits ripen well after harvesting, but if they are harvested too unripe, the milky latex coming out of the cut stem spoils the appearance of the fruit. For proper marketing the fruits are picked (*i*) when the fruits start softening on the tree, and (*ii*) when the colour below the dark brown rind which changes into light hue at which stage the latex also stops oozing out of scratching. After picking, the fruits are spread on mat under shade for sometime to allow the drying of latex, if any. Thereafter, the fruit are packed in the baskets having padding of banana leaves, rice straw or soft grass.

Karmakar and Joshi (1940) reported that fruits ripen satisfactorily at 12 and 13.5°C but cannot do so at 7°C. Under ordinary conditions it can remain healthy for 5 weeks though ripe fruits can be kept in cold storage for about 6 weeks at 0° or 2°C. According to Singh and Mathur (1954) they could keep full grown hard fruits for eight weeks at 2-3°C but subsequently they could not be kept for more than five days. They further noticed a rapid decrease in ascorbic acid content during storage. Sapora fruit is highly astringent when it is unripe, due to the presence of polyphenols. At harvest, sapora fruit contains about 2.4% polyphenols, which on ripening decreases to 1.8%. On ripening they emit a sweet smell and become soft. Principal chemical substances are carbohydrates and free sugar, tannins are also present. Starch is absent or present in small quantities (Joshi and Paralkar, 1991). Owing to their richness in pectin the fruit is suitable for making jelly and Jam. Lakshmana and Reddy (1999) studied the effect of $CaCl_2$ and $Ca(NO_3)_2$ on the post-harvest quality of sapota cv. Kalipatti fruits. It was observed that when fruits of uniform maturity were harvested 3 weeks after treatment and stored for two weeks, the fruits treated with 1% $CaCl_2$ showed significantly higher total sugar content.

Meza and Manzano (2000) observed the effect of 0.75% alum and 0.5g benomyl/litre on storage for 20 days. Fruits deteriorated after 6 days at 28°C. Storage at 10°C gave the best results.

Many varieties of sapota are known in different parts of India. These are divided into two main groups, one having round fruits and the other oval fruits. However, this character is not stable. The important varieties are Kalipatti and Chhattri in the Mumbai area, Pala, Dwarapudi and Kirthabarthi in Andhra Pradesh and Cricket ball in Tamilnadu and Bengal.

6.2. INSECT PESTS

6.2.1. Stem Borer : *Indarbela tetraonis* Moore

(Family-Metarbelidae : Order-Lepidoptera)

The larvae are wood borers and they are confined to tropical region. It attacks a large number of plants like, mango, guava, ber, citrus, orange, jamun, aonla etc.

The larva bores into stem and feeds on bark, only one larva is seen inside a bore hole which is used only as a shelter when it does not feed during day time. The incidence of the pest is easily made out by the presence of the peculiar winding gallaries, generally near the forks or angles on the stem and branches.

Control

1. By injection of 5 ml of kerosene.
2. Use of Carbon disulphide (CS_2), endrin (0.04 per cent).
3. Use of B.H.C. (0.2 per cent) or DDT (0.5 per cent) into the hole and sealing it with cotton, wool and mud gives good control.

6.2.2. Scale Insect : *Pulvinaria psidii* Mask.

(Family-Coccidae : Order-Hemiptera)

P.psidii is the most common scale insect in South India. It infests guava, mango, sapota, tea, coffee, *Ficus* etc.

The females are flat and elongate-oval with the integument smooth or covered with wax. They may possess legs or legs may be absent. Antennae are much reduced. The males may be alate or apterous.

6.2.3. Brown Scale : *Saissetia hemisphaerica* Ber

(Family-Coccidae or Lacaniidae : Order-Hemiptera)

S. hemisphaerica known as brown scale is a pest of coffee, citrus, sapota guava, mango etc.

Control

Spraying parathion (upto 0.05%) and malathion (0.04%) may be effective in controlling the insect.

6.2.4. Sapota Leaf Webber : *Nephopteryx eugraphella*

(Family-Phycitidae : Order-Lepidoptera)

This is an important pest on sapota. The larvae constructs a tunnel of webs and grass and feeds on the green matter of leaves. All plant parts like leaf buds, flower buds, tender leaves and fruits are attacked. Occasionally three or four larvae may be seen in a tunnel.

Apart from sapota this moth is known to infest cured tobacco and *Mimusops elengi*. Moth is grey coloured and lays pale yellow oval eggs in groups of four or three on leaves. A female can lay 226 eggs in seven days. Full grown larva is 25 mm long, slender and pinkish with few longitudinal close-sett lines on the dorsal surface. It pupates in the leaf web. Total life cycle is 32 to 45 days.

Control

1. Removal and burning of infested plant parts.

2. Spraying of BHC (0.1%) may be beneficial.
3. Spraying of Calcium arsenate (Singh, 1969).

6.2.5. The Castor Aleyrodid : *Trialeurodes ricini* Misra

(Family-Aleyrodidae : Order-Hemiptera)

This is a pest of Castor plant (*Ricinus communis* L.) It occurs throughout India.

The adults are pale yellow with white wings covered with waxy powder. The female prefers tender leaves for oviposition. The eggs are laid in groups or are scattered on lower surface of leaves, the nymphs are light, transluscent yellow in colour and entire margin encircled by a broad fringe of closely set thick waxen filaments. The nymphs suck the sap from the under surface of leaves in large numbers as a result of which the leaves show yellow patches. Honey dew secretion favours growth of sooty mould (*Aspergillus niger* van Tiegh.). In young plants severe infection results in gradual drying of leaves and ultimate death of the plants. *T. rara* Singh occurs along with the species in a predominant way.

It occurs on Ricinus communis, Achras sapota, Breynia rhamnoides.

Natural enemies

The drosophilid fly *Acletoxenus indica,* Coccinellids—*Cryotpgnatha flavescens* and the lygacid bug *Geocoris tricolor* are predaceous on the nymphs of Aleyrodid. *Prospaltella* is a parasite. In Coimbatore (T.N.) the fungi *Spicaria* sp. And *Cladosporium* sp. are parasites.

Control

Use of Parathion (0.02%) or Carbophenophion (0.06%) is recommended to control the pest effectively.

6.2.6. Leaf folder : *Banisia myrsusales elleralis* Walker

This pest has been reported to infest sapota in Gujarat and Punjab (Sandhu and Saran 1980; Patel and Jhala 1991). The first instar tiny larva nibbles mostly near the midrib of tender leaves, while second, third and fourth instar larvae fold the leaf from midrib and web it from margin giving an appearance of pea pod. The larva remains inside the leaf and feeds on green matter.

Control

The pest is active during Nov.–Dec. in Gujarat (Patel and Jhala, 1991) for its love for cold and dry days and therefore, control measures should be concentrated during this period. Spray of Monocrtophos 0.036% or Fenthion 0.1% or Endosulphon 0.07% is recommended (Chundawat, 1998).

6.2.7. Leaf miner : *Acrocercops gamoniella* Staiton

Fletcher (1920) reported this insect from Bihar. In Gujarat it has been reported to infest sapota (Patel 1981). The tiny larvae damage the tender leaves. This pest affect sapota throughout the year.

Control

To control this insect spray of malathion 0.05% or dichlorovos 0.05% or dimethoate 0.05% are recommended (Bhutani and Srivastava 1983).

6.2.8. Table. Some More Insect Pests of Achras sapota

Sr. No.	Name	Family	Order
1.	*Idioscopus atkinsoni (Mango-hopper)*	Cicadellidae	Hemiptera
2.	*I. clypealis (See 1.2.1)*	Cicadellidae	Hemiptera
3.	*I. niveosparsus (See 1.2.1)*	Cicadellidae	Hemiptera
4.	Walkeriana cinerea Gr.	Margarodidae	Hemiptera
5.	*Flaka ocellata Fb. (Green plant hopper)*	Flatidae	Hemiptera
6.	*Frankliniella dampfi Priesner (Flower thrips)*	Thripidae	Thysanoptera
7.	*Acrobasis romonella (Bud borer)*	Phycitidae	Lepidoptera
8.	*Virachola isocrates Fb. (fruit borer)*	Lycaenidae	Diptera
9.	*Dacus dorsalis Hendel (fruit fly) (See 1.2.2)*	Tephritidae	Diptera
10.	*D. zonatus Saund. (fruit fly)*	Tephritidae	Diptera

DISEASES CAUSED BY FUNGI

6.3. SOOTY MOULD

Sooty mould caused by *Capnodium* sp. is one of the common diseases foundon many fruit trees. This ectoparasitic disease is most common on chiku, causing considerable harm to the normal development of the plant (Rangaswami, 1984).

6.4. ROOT ROT

Causal organism : *Ganoderma lucidum*

Ganoderma root rot is a problem in old sapota plantation, where the lateral roots become discoloured with whitish mycelial strands growing all around. The bark of roots become loose and scaled off. The affected roots are very light in weight. Later these strands turn dark and the fungus appear at the base of the trunk and in rainy season bracket like fructification appear. Disease can be controlled by digging the trench around the trees, and drenching with suitable fungicide like Bavistin 0.1% (Rawal 1991)

6.5. PHAVO

Studies were undertaken in sapota (*Manilkara zapota*) cv. Cricket ball to determine the critical period of disease development and the effectiveness of fungicides and mulching, alone and in combination for disease control (Rawal 1998).

6.6. PHAVOPHLEOSPORA LEAF SPOT

A leaf spot disease caused by *Phavophleospora indica* Chinnappa is common in many parts of South India.

The disease starts as yellow coloured dots few mm in diameter, which gradually increase in size and are surrounded by violet coloured ring, this becomes brown coloured later on.

6.7. LASIODIPLODIA ROT

The disease is caused by *Botryodiplodia theobromae* Pat. (Tandon, 1967). The disease starts as a water soaked spot at the stem end. It enlarges gradually and extends towards the stylar end and within 8-10 days, the fruits become completely rotton. At that stage pycnidial stroma becomes evident on the fruit surface. This disease is characterized by the production of soft watery rot. Completely rotton fruits become shapeless.

B. theobromae and *Pestalotia sapotae* have been isolated from the fruits from Jaipur, Jodhpur, Delhi, Meerut, Kanpur, Lucknow, Allahabad, Chennai, Cape Camorin, Hyderabad, Mumbai, Bhusaval and Kolkata markets.

Free amino acids present in the healthy and infected tissues was analysed by Tandon (1967). He found that except for serine and glycine most of the other aminoacids were present in lesser quantity in infected fruits.

Tandon (1967) studied the effect of temperature on tropical fruits. He found that sapota fruits stored at low temperature (10°C) did not develop any rot for 8 days. Srivastava (1966) working with tropical fruits found that sapodilla fruits inoculated with *B. theobromae* escaped rotting for ten days at 10°C.

6.8. PESTALOTIA ROT

Pestalotia sapotae P. Henn. was found responsible for this rot.

The disease is characterized by the appearance of light brown spots on the surface of the fruits. Gradually they are covered with white fluffy mycelial growth. At a later stage acervuli, oozing dark green viscid masses of spores, appear on such lesions. The pathogen causes moderately soft rot. In the initial stage the infection is not clearly distinguishable from the healthy fruits because of their colour.

Changes brought about in host tissues during pathogenesis were studied at Allahabad (Tandon, 1967). It was observed that the *P. sapotae* and *Botryodiplodia theobromae* were slow in hydrolysing sucrose as traces of this disaccharide were detected upto 10th and 12th day respectively. Both the organisms simultaneously consumed glucose and fructose.

Five organic acids were present in the healthy fruit. Malic and citric acids were in greater quantity, whereas succinic and oxalic acids were in traces. Other acid was fumaric acid. Only three citric, malic and fumaric acids were detected in infected fruits.

Effect of temperature on the fungal decay of sapota fruits was studied by Tandon (1967). In case of *P. sapotae*, there was no rot at 10° and 15°C upto 8 days of incubation but in case of *B. theobromae* 2.8% rot was observed after 8 days at 15°C. Fruits infected with *P. sapotae* did not show any rot at 20°C upto 4th day but after 8 days of incubation, the rot was 20%. The loss increased between 25° and 35°C. *B. theobromae* was capable of causing rot at 15°C. Fruits inoculated with both the organisms when brought back to 25°C after incubation at 10° to 20°C for 4 days showed considerable loss. The rot was maximum at 30°C.

H—Healthy

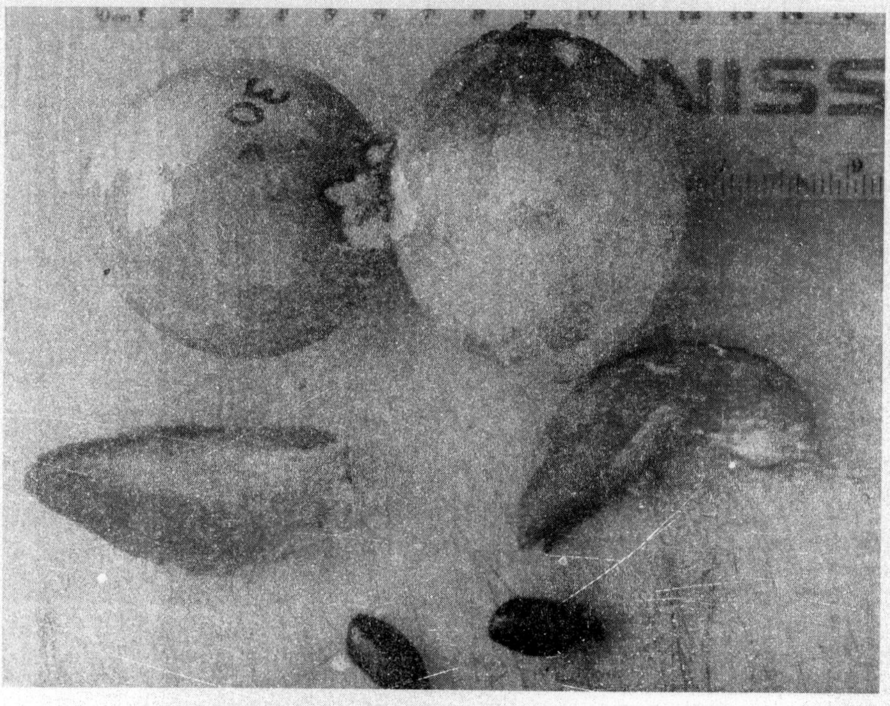

D—Diseased

Fig. 32. Pestalotia rot of chiku. H—Healthy, D—Diseased.

6.9. HENDERSONULA ROT

The causal organism of this disease is *H. toruloidea* (Tandon, 1967). The rotten portion are soft, wrinkled and shunken. In advanced stages liquid ooze out from the infected region.

6.10. ASPERGILLUS ROT

It is caused by *Aspergillus niger* van Tieghem.

The rot starts as a water soaked lesion. It enlarges with age and gets depressed in the centre. In later stages black mouldy growth of *A. niger* appears on the infected surface of the fruit. The whole fruit becomes infected with black pin head like conidial mass which develop well between 10 to 12 days.

Arya and Mathew (1990) reported control of this rot by six medicinal plants. Extracts of *Orobanchae aegyptiaca* Pers was most effective. It gave 52.9% spore inhibition at 75% concentration. 43.3 and 41.1% spore inhibition was obtained by *Eucalyptus occidentalis* Skeels and *Cuscuta reflexa* Roxb. respectively.

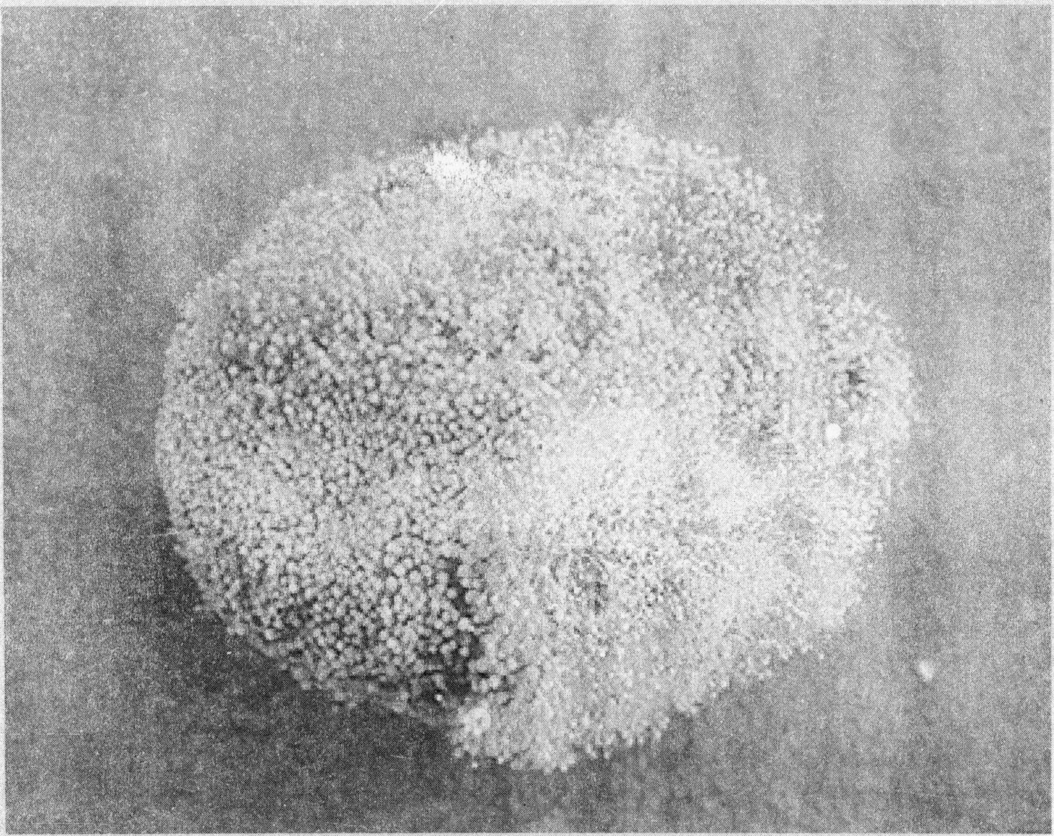

Fig. 33. Aspergillus rot.

6.11. EMERICELLA ROT

The disease starts as a water soaked lesion, which enlarges with age and got depressed in the centre. Fruits turn green due to the presence of green conidial heads, which showed the presence of rounded fruiting bodies (Cleistothecia). Arya (1995) reported this disease from the chiku orchards around Baroda. Imperfect state of *E. nidulans* was earlier reported from mango (Goyal *et al.*, 1971) and Bel (Sharma, 1953).

6.12. PESTALOTIOPSIS ROT

Causal organisms *Pestalotiopsis glandicola* (Cast.) Steyaert

Dhingra *et al.* (1980)

P. neglecta (Thom) Stey.

Jamaluddin and Tandon (1977)

P. mangiferae (P. Henn.) Stey.

Jain *et al.* (1981)

P. palmarum (Cooke) Stey.

Arya (1995)

P. versicolor Stey.

Gupta and Sehgal (1974)

During a local survey of diseases of fruits in 1972 and 1973 in Allahabad (Jamaluddin and Tandon 1977) *Pestalotiopsis neglecta* was found on chiku fruits. About 4 to 7% rotted fruits were recorded in transport boxes as well as in marketing processes. In local orchards around Baroda, Arya (1995) observed prevalence of another species *i.e. P. palmarum* on chiku fruits. This species has been earlier recorded from *Zizyphus* and kamarkh.

The infection first appeared as a soft rot, water soaked, coloured area, not confined to any particular point on the fruits but injury was necessary for the initiation of the disease. Later the diseased area enlarged, showed black dot like acervuli of the fungus.

Gradually the entire fruits rotted in 8 days.

6.13. PHOMOPSIS SOFT ROT

Lal and Rai (1980) reported a new soft rot of chiku fruit in the months of November-March. From such infected fruits they reported a new species of *Phomopsis, i.e. P. sapotae* Lal and Rai.

The infection starts from the stalk end and the disease develops as a water soaked spot which extends and covers the whole fruit causing soft rot. The fungus penetrates the mesocarp turning its colour to dark brown. The diseased portion develops white mycelial growth and after 7-10 days of infection a number of black pycnidia are produced all over the surface of the fruit. The infected fruits shrink and get deformed.

Fig. 34. Gloeosporium leaf spot of chiku.

Host Range

The fruits of *Psidium guajava* L., *Vitis vinifera* L., *Musa paradisiaca* L., *Mangifera indica* L., *Zizyphus mauritiana* Lamk. etc. are infected by the pathogen.

Control

Treating the plucked fruits of chiku with 1000 ppm Bavistin controls the disease.

6.14. TABLE : SOME MORE FUNGI REPORTED FROM SAPODILLA

Sr. No.	Organisms	Nature of Infection	Localitie	Reference
1.	*Gloeosporium rubi. Ellet Ev*	Leaf spot	Allahabad	Srivastava *et al. 1964a.*
2.	*Nigrospora oryzae (Berk. et Br.) Petch*	"	"	"
3.	*Chaetomium* sp.	Leaf spot	Pune	Tandon, 1967
4.	*Pestalotia sapotae P.Henn.*	Leaf spot fruit & rot	Bhusaval Allahabad Meerut	"
5.	*Bortyodiplodia theobromae Pat.*	Leaf spot, fruit & rot	Bhusaval Allahabad Meerut	Tandon, 1967
6.	*Fusarium roseum Link*	Fruit rot	Delhi	"
7.	*Botryosphaeria dothedia Moug. ex Fr.*	Fruit rot	"	Jain, *et al.* 1981
8.	*Curvularia lunata (Wakker)* Boedijn var. *aeria*	Leaf spot	"	Kore & Bhide, 1977

Fig. 35. Phomosis soft rot of chiku.

7 Papaya

7.1. GENERAL ACCOUNT

Carica papaya L. is commonly known as melon tree, papaya, papaw or papita. It belongs to the family Caricaceae. Spanish and Portugese sailors were responsible for the dissemination of papaya to the tropical and subtropical countries. It has been introduced into China as an Indian plant, even though this tropical plant is a native of Mexico. It came to India via Malasia in the 16th century. It has gained the greatest popularity in Hawaii. It is fifth most important fruit of the country.

It is rich in enzyme **papain,** which helps in the digestion of proteins. A small piece of raw papaya cooked along with meat makes it very tender. The papain content of the fruit decreases rapidly as it ripens. Papain is a very useful product. It is said to remove skin blemishes, is considered beneficial for the treatment of stomach ulcers, diphtheria and even cancer. It has also industrial uses. It is used in the clarification of beer, in the tanning industry and in the manufacture of Chewing gum. Raw papaya is cooked as a vegetable or made into pickles. The ripe fruit and the seeds are also considered to have medicinal properties against disorders of liver, spleen and the digestive tract. According to Singh (1969), the papain extracted from unripe fruit can fetch Rs. 35,000 per hectare. After the extraction of papain the scarred fruits are harvested and can fetch another Rs. 15,000 per hectare. The seeds and unripe fruit are rich in benzyl isothiocyanate, a sulphur-containing chemical that has been reported to be an effective germicide and insecticide (Mitra, 1997). Carpaine, an alkaloid found in papaya leaves, has also been used for medicinal purpose (Olaya, 1985).

It is a nutritive fruit containing 85-90% water, 10-13% sugar and 0.6% protein and an equivalent amount of minerals, consisting mainly of iron, calcium and phosphorus. It has 2020 international units (mango has 4800 units) of Vit. A, 'Carotene', 70 mg of ascorbic acid per 100 g of fruit.

The colour of the pulp turns yellow or red during ripening. The carotenoids (Provitamin A) content estimated as β carotene showed a five- to tenfold increase in yellow-fleshed cultivars between the mature green stage and the full ripe stage (Selvaraj *et al.*, 1982). In red fleshed cultivars, however, the change in colour is associated with a marked increase in lycopene content. The major difference between the yellow and red fleshed papaya is the complete absence of lycopene in the yellow-fleshed types (Chan, 1983). Linalool has been found as the major volatile compound in papaya. Flath and Forrey (1977) identified as many as 106 volatile compounds in papaya. Fruits of cv. 'Solo' has a high percentage of Linalool (up to 94%), as translinalool oxide. The relative proportions of different volatile components vary with the stage of ripeness; a merely 400 fold increase of linalool and a seven fold increase of benzyl isothiocyanate production was found in Solo papaya during ripening (Flath *et al.*, 1990). The volatile compound methyl butanoate is considered responsible for the sweetish odour quality of papaya.

Papaya is a tropical plant, but also does very well in a mild tropical climate. It is very sensitive to frost, it does not like strong, hot or dry winds. Mountain papaya (*C. candamarcensis*) can grow up to an elevation of 2,000 metres, but it produces many fruits of poor quality. In high rainfall areas and poor drainage of soil, the cultivation of papaya is impossible due to death of the plants caused by the Collar-rot disease (see Article 7.4). However, a hardy species (*C. monoica*) grows in the upper Amazon basin with an annual rainfall of 375 cm.

Papaya is grown all over India. The total area under papaya in India is estimated at about 10,000 hac. Of this the major part about 4,000 hac., is in Bihar. Assam has slightly less than this area. U.P., Andhra Pradesh, Gujarat, Tamil Nadu and Karnataka are other states where papaya is grown.

The Date palm has male and female trees, but the papaya has male, female, hermaphrodite (with male and female flowers on the same tree) and some other complex forms. The sex of the flowers on the same tree can undergo change with season, age and injury. The male flowers hang in long trusses. It has been claimed that the sex of the male plants can be changed by measuring or decapitating them. None of these methods has been found practicable and some of them are baseless. However, the decapitation of the plant, if only one new shoot is retained on it; gives fruits of good quality in Florida. If the female trees are pollinated with hermaphrodite ones the progeny consists only female and hermaphrodite plants. In this case, male plants which do not bear any fruit, are not required for pollination of females.

In Central India, papaya flowers throughout the year and fruit can be had any time during the year except the monsoon months (July to October). Seeds are sown in rainy season. Transplanting in December gives fruits on plants when they are much shorter. Papaya is often planted as a filter tree between newly planted mango and orange grooves. It is planted two to three metres apart. In Jaipur city (Rajasthan) papaya can be seen growing in small fields enclosed in three metre high stone walls.

Papaya is a shallow-rooted plant, which should never be given deep tillage. Irrigation should be given after every 10 days in winter and every week in summer, provided there is no rain. Irrigation should be done by the ring method, which is followed in Gujarat and prevents the water from coming in contact with the stem, thus reducing the possibilities of Collar-rot.

The fruits are picked when they are firm and only the apical end has started turning yellow and the latex is no longer milky. The yield per tree in India varies from 20 to 150 fruits per annum. The average weight of a fruit should be at least 1 kg.

Since papaya is propagated from seed, which is not produced under controlled conditions, the identity of different varieties is not clear. Solo is the well known variety in Hawaii islands. Some other varieties are Hawaii, Singapore and Sri Lanka. In our country the main vareities are Washington, Coorg, Honeydew, Madhu, Bindu, Ranchi and Sri Lanka. Co-1 is a selection made in Tamil Nadu from the variety Ranchi from Bihar. The Gujarat variety bears heavily, but its fruit is not of good quality. The variety Barwani from Indore is of good quality.

The introduction of Hawaiian lines (Solo) having small sized fruits, has revolutionized its cultivation. Development of new varieties like Surya is tried by IIHR, Bangalore. The papaya Surya is a cross between Sunrise Solo and Pink Flesh Sweet. It is gynodioecious and provides medium sized fruits, weighing on an average 600-800 g each with red pulp. Punjab Sweet is deep yellow with a melting texture and excellent flavour. The average yield is 50 kg. It is comparatively better tolerant to frost than Pusa Delicious, grown presently.

7.2. BIRD AND INSECT PESTS

Birds are the most serious pests of papaya. Bird-scaring is an important item of expenditure in the cultivation of this crop.

The most important insect is red spider mite, which can be controlled by spraying lime sulphur. The caterpillar, *Dasyses rugosellus*, and root knot nematodes have also been found.

7.2.1. Aphids : *Myzus persicae Sulz.*

(Family-Aphidiae : Order-Hemiptera)

Aphids are small bodied delicate, pear shaped insects that infest the stem, petiole or leaves of variety of plants in large numbers and suck up the sap causing serious damage. They affect young tender parts. It is a vector of Papaya mosaic. (7.3.1.). This disease may be spread by some other aphids.

The characteristic feature of an aphid is its pear shaped structure, fairly long antennae, a pair of compound eyes, several joined rostrum, two-segmented tarsi with paired claws, nine pairs of lateral spiracles and pair of cornicles or 'honey tubes' The cornicles secrete a waxy fluid which affords protection from predaceous insects.

7.2.2. Cotton White Fly : *Bemisia tabaci Gennadius*

(Family-Aleyrodidae : Order Hemiptera)

Distributed throughout India, Sri Lanka, Nigeria, Congo, West Africa, it is a vector of leaf curl virus on tobacco, cassava, cotton, papaya (7.3.2) etc. It is a polyphagous species.

The female lays the eggs on the undersurface of leaves. Oviposition periods ranges from 2 to 12 days. Incubation period ranges from 3 to 33 days. A complete life cycle from egg to adult may occupy 14 to 107 days. White flies are small inconspicuous insects restricted to the tropics. All species live on the leaves of plants, mostly on the under surface. The aleyrodids are small insects with seven segmented antennae, two ocelli and a pair of reniform compound eyes. Wings are of equal consistency, opaque, whitish clouded or mottled with pots or bands.

Host range

It has a large number of host plants like Cotton (*Gossypium* sp.), *Brassica* spp., tobacco (*Nicotiana tabacum*), *Solanum* spp., *Phyllanthus* sp. etc.

Control

Sprays of malathion (0.1%), phosphamidon (0.1%), dimethoate (0.03%), barrathion (0.025%), fenitrothion (0.05%), methyldemeton (0.025%), carbaryl (0.1%) can effectively check the spread of insect pest.

7.3. VIRAL DISEASES

7.3.1. Papaya Mosaic

This is a devastating viral disease prevalent in the central belt of India. It is reported from East Aftica, Rhodesia, West Indies, Hawaii, Cuba, Brazil, Purerto Rico and U.S.A. In India it was first reported from Mumbai and Pune in 1958 by Capoor and Varma and was the main cause for abandoning large scale commercial cultivation of papaya in some parts of Deccan (Rangaswami, 1984).

The disease is most serious on young plants. The top young leaves of the diseased plant are much reduced in size and show blister like patches of dark green tissue, alternating with yellowish-green lamina and puckering. The leaf petiole is reduced in length and the top leaves assume an upright position. The fruits on diseased plants develop circular water-soaked lesions with a central solid spot. The fruits are elongated and reduced in size, they do not attain full maturity and are spoiled being exposed to the sun.

The virus is spread in planations by at least 6 aphid species mainly *Myzus persicae* Sulz. None of these colonise papaya. It is not seed transmitted and is thermolabile and non-persistent.

All measures adopted for controlling the disease failed. The only possibility of its control lies in breeding a resistant variety. One species of *Carica, C. cauliflora* is immune to the mosaic virus. Possibility is being explored to transfer the gene for resistance from *C. cauliflora* to *C. papaya* and to grow the resulting hybrid by the embryo culture method. According to Bhargava and Khurana (1969), weekly spray with 1% ground nut oil helps to prevent infection.

7.3.2. Leaf Curl of Papaya

The disease was commonly observed in Kitchen gardens and commercial orchards around Delhi (Mehrotra, 1985). The disease has also been reported from Bihar, U.P. and Gujarat.

Crinkling and curling of the leaves accompanied by vein clearing and reduction in size of the leaves are the marked symptoms. Leaves become leathery and brittle and the interveinal areas are raised on the upper surface due to hypertrophy other symptom is the downward and inward rolling of dark green areas. Sometimes all the leaves at the top of the plant get affected by these symptoms. The petioles are twisted in a zigzag manner. In the advanced stages of the disease, defoliation takes place and plant growth is stunted.

Tobacco virus 16 or Nicotiana virus 10 is the virus responsible for leaf curl of papaya. The virus is not mechanically transmitted (Nariani, 1956). It can be readily transmitted by grafting or juice inoculation. In nature the most important agent of its transmission is the White fly *Bemisia tabaci* (Nariani, 1956). Virus also causes infection to tomato, tobacco, *Zinnia* and hollyhock (*Althaea*).

Control

To control leaf curl is difficult. It is recommended that the diseased plants should be cut down in early stages of disease development. Tomato, tobacco, *Zinnia* or *Althaea* should not be grown in the close vicinity of papaya plantation, and a strict watch should be kept on the disease incidence in papaya nursery as well as in the young papaya plantation.

The control measures if adopted by all the growers of a particular zone simultaneously, will give good results. Contrary to it, if efforts are made to control the disease in individual holding, it will certainly amount to loss of energy and waste of efforts.

7.3.3. Papaya ring spot virus

Studies on transmission of papaya ring spot virus were conducted by Taya *et al.* (1999). Relationship between PRSV and *A. gossypii* Glover is established. Unstarved adults transmitted PRSV to 10% of papaya plants and nymphs failed to transmit PRSV.

DISEASES CAUSED BY FUNGI

7.4. FOOT AND STEM ROT OF PAPAYA

Foot rot of papaya is caused by *Pythium aphanidermatum* (Eds.) Fitz. This fungal disease usually appears in rainy season and its severity depends upon the intensity of rainfall. This pathogen also causes stem rot and damping- off seedlings. In addition, *Phytophthora parasitica* var. *nictoianae, Rhizoctonia solani* and *Macrophomina phaseoli* (*R. bataticola*) are also associated either singly or in combination with foot rot in different regions of the country.

It is characterized by the appearance of spongy, watersoaked patches on the stem near the ground level. These patches enlarge rapidly and girdle the stem, causing rotting of the tissues, which then turn dark brown or black. Under wind pressure the plants may topple over and die. In less severe infections one side of the stem may rot, the plant may be stunted, the fruits shrivelled and drop and finally there is slow death of the plant. Occasionally the terminal leaves begin to droop, yellow, wilt and fall. Perpetuation is by thick-walled oospores.

Control

Use well drained soil. Avoid water-logging. Scarp affected plants and paint with Bordeaux paste. Sprays of Bordeaux mixture (6 : 6 : 50) with alboleum as sticker and application of Vapam (Sodium methyl dithicarbomate) drench to soil has been found fairly effective in Uttar Pradesh (Raychaudhuri and Lele, 1967).

The damping-off phase of the disease can be effectively controlled by soil treatment with Esso fungicide 40.6 (0.2%) before and during emergence (Tandon, 1959). Soil application of Brassicol, Captan or Formaldehyde solution minimizes its incidence. Seed treatment with Agrosan GN or Ceresan dry, greatly improves emergence. Both seed and soil treatments are recommended for effective control.

Improvement of aeration of soil and avoidance of water coming into direct contact with the stem prevent disease development. Raising of nursery in sterilized beds is also helpful.

7.5. ASPERISPORIUM LEAF SPOT

It is caused by *Asperisporium caricae.* It was first reported from Rhodesia (now Zimbabwe by Rothwell, 1975).

It produces numerous black velvety spots on the lower surface of the leaves with necrotic yellowish to white upper surfaces. It is more severe on lower older leaves during summer months in Karnataka and Tamil Nadu states and results in leaf drying. Sporodochia are hypophyllous dark brown to black.

No control measures are known at present against this disease.

7.6. ASCOCHYTA LEAF SPOT

It is caused by *Ascochyta caricae papayae.*

This disease was reported to be serious on var. Coorg Honey Dew in Coorg area (Karnataka state) by Ullasa *et al.* (1974). Both sexual and asexual stages were encountered simultaneously on the diseased leaves in the field.

Control

Spraying with Dithane Z-78 (0.2%) gave effective control.

7.7. POWDERY MILDEW

The disease is caused by *Sphaerotheca fuliginea* (Schlecht. ex Fr.) Poll (Munjal and Kapoor, 1973) and *S. caricae- papayae* (Tanda and Braun, 1985). Its conidial state is already known as *Oidium caricae* Noack. The disease is serious in nursery and on young plants.

The mildew develops as small circular patches on both sides of the leaves and stem of the young seedlings of *Cucurbita pepo* and *Carica papaya.* The patches gradually extend, coalesce and cover the entire leaf surface. Badly infected leaves curl, dry up, hang down and ultimately fall off. When large number of leaves are involved, the entire plants or seedling may die. Death is more frequent in older plants. Plant collapses giving damping off appearance. In the young papaya plants the infection remains confined to young succulent tissues like leaves, pedicels etc., and does not spread to stem.

In IARI, New Delhi, infection first occurred on *Cucurbita pepo* and later on seedlings of *Carica papaya* inside the glass house. Cleistothecia appeared profusely both on leaves and stems of *Cucurbita.* But on papaya were scanty on stems only, apparently because the infection on seedlings started much later and these died off quickly. Cross inoculation were made with conidia from either host to healthy plants. The Mildew appeared on both the hosts in 5-7 days after inoculation at 24—30°C, which confirmed that the fungus infecting the two hosts is the same.

Control

Spraying with Sulfex (0.2%) or Bavistin (0.1%) or NF 44 (0.1%) at fortnightly to monthly intervals was found effective at Bangalore (Ranwal *et al.*, 1981). Proper spacing minimises the disease.

7.8. ANTHRACNOSE OR COLLETOTRICHUM ROT

Three spp. of *Colletotrichum* viz., *C. papayae* Sacc. (Wangikar and Raut, 1972), *C. dematium* (Pers. ex Fr.) Grove (Solangi & Malik, 1971), *C. capsici* Syd. Butler & Bisby (Lal *et al.*, 1979) and *C. gloeosporioides* Penz. (Srivastava *et al.*, 1964; Prasad & Verma, 1970) are known to cause anthracnose of papaya.

C. capsici. Characteristic symptoms of an anthracnose were noticed when *C. capsici* was artificially inoculated on freshly harvested healthy ripe fruits of papaya. The disease started as water-soaked lesion which subsequently progressed in a circular manner. Six days after incubation, soft rot spot of 4 to 5 cm. diam. was recorded with brown mycelium. On this discoloured area numerous acervuli of fungus were present.

Dastur (1921) had earlier recorded occurrence of this fungus on papaya flowers. Later on Deighton (1936) had reported *C. capsici* from papaya fruit in Sierra Leone (Africa).

C. dematium. *C. dematium* is the main cause of anthracnose of papaya in the Hyderabad and Karachi areas. Physiological studies revealed best growth at 28°C and pH 6.5 on sucrose nitrate agar and Richard's solution with KNO_3 and Maltose. Setae and sporulation were maximum in constant light and decreased in darkness. Of the fungicides tested in lab Polyram ultra completely checked the growth of the fungus.

C. papayae. The spots appear as brown, superficial discoloration of the skin and then they develop into circular, slightly sunken spots, 1-3 cm in diameter. Usually such spots appear water-soaked. Gradually the spots coalesce and form bigger spots. Under humid conditions an incrustation of salmon pink spores, often arranged in a concentric pattern, may develop on the surface of some of the older spots. *C. papayae* was found responsible for this disease.

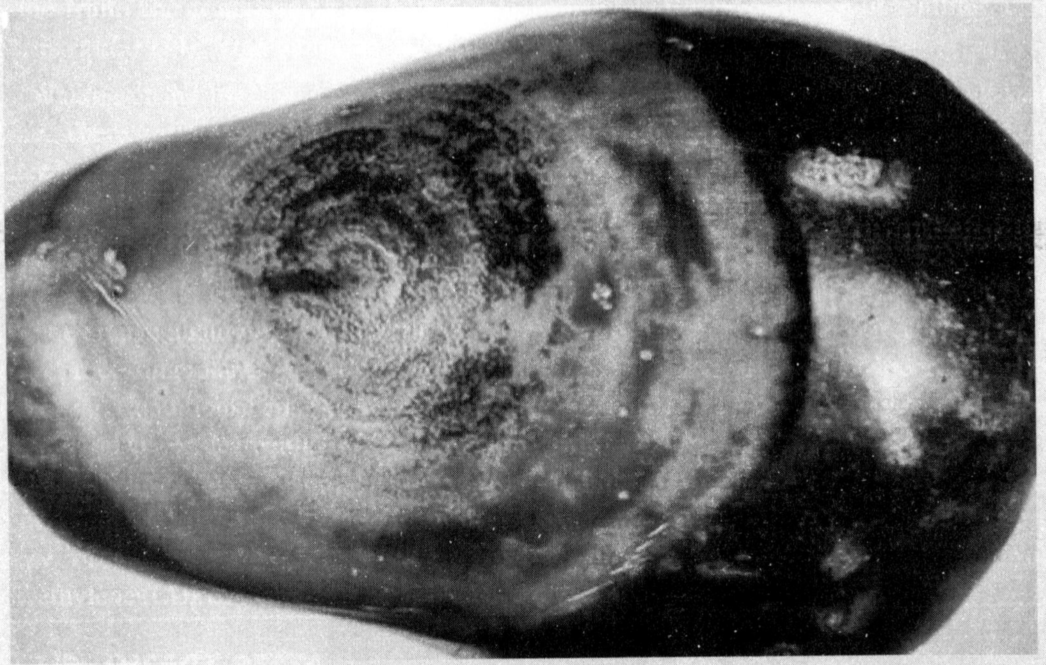

Fig. 36. Rot of papaya caused by *Colletotrichum capsici*.

C. gloeosporioides. Pathogen caused leaf spot in Allahabad (Srivastava *et al*., 1964a) and fruit rot in Arrah, Bhagalpur, Muzaffarnagar, Patna and Ranchi in Bihar (Prasad and Verma, 1970).

Diseased lesions appeared as pin incrustations.

7.9. GLOEOSPORIUM ROT

Cheema and Dani (1930) as well as Naik (1949) reported the occurrence of anthracnose (caused by *Gloeosporium* and *Colletotrichum*) from different parts of the country. They described the symptoms and suggested certain control measures. Grewal (1954) and Tandon (1967) described fruit rots of papaya caused by *G. papayae* and *C. papayae* from Allahabad.

Gloeosporium papayae P. Henn. causes leaf spot and fruit rot diseases referred as anthracnose along with *Colletotrichum* spp. The symptoms on fruit are characterised by water-soaked lesions which enlarge and after few days watery oozing of fungal spores occurs.

Control

The disease can be controlled by spraying Bordeaux mixture on tree and dipping the picked ripe fruit in 2000 ppm Bavistin.

7.10. STALK ROT OF PAPAYA

Fusarium solani (Mart.) Appel and Wr. was repeatedly isolated from dry rotten stalks of papaya fruits, while they were on the plants. The disease caused premature fruit fall and was quite common wherever the plants are grown (Shukla *et al.*, 1978).

The symptom first appeared as pale yellow discolouration towards plant side of the fruit stalks. Later on, the stalk bent at the infection site and started drying. The stalk was completely dry and shrivelled as the disease advanced and became light brown in colour. Due to formation of abscission layer, fruits fall prematurely.

Control

Efficacy of three fungicides (Tecto-40, a formulation of Thiabendazole, Benlate and Dithane Z-78) was tested in artificially inoculated plants. Tecto-40 at 1000 ppm and Benlate at 1500 ppm were found to control the rot completely, while Dithane Z-78 was partially effective.

7.11. ASCOCHYTA ROT

Tandon (1967) isolated *Ascochyta caricae* from papaya fruits from Ranchi and Patna. But Prasad *et al.* (1970) denied its occurrence. The organism was reported from Assam by Chowdhury (1950).

The initiation of the disease is evident by the appearance of water soaked lesions which blackened and produced numerous pycnidial bodies in advanced stages.

7.12. ULOCLADIUM ROT

A new soft rot of papaya caused by *Ulocladium chartarum* (Preuss) Simmons was reported by Lal *et al.* (1984).

The disease was observed in months of March and April, 1983. When healthy fruits were artificially inoculated with the pathogen, small rounded to oval spots developed which increased gradually. In advanced stage of rotting, superficial mycelium covered the infected portion and after 10 days almost whole fruit was rotten. Injury was found essential for the development of disease.

Control

Treating the fruit with 1000 ppm Bavistin or 1250 ppm Tecto-40, reduces the loss.

7.13. HYALODENDRON ROT

The disease caused by *Hyalodendron* sp. (Arya *et al.*, 1986) was characterized by the appearance of white to light brown coloured, small irregular necrotic lesions over the ripe fruit. The infection was severe in case of ripe fruits and resulted into 25-30% loss during storage and marketing. The disease was highly favoured by damp conditions of storage.

The infection was also noticed simultaneously occurring along with Fusarium rot of papaya but the typical brownish necrotic spots were characteristic of the disease.

7.14. RHIZOPUS ROT OR WATERY FRUIT ROT

Rhizopus rot is caused by *R. nigricans* Ehrenb. Srivastava *et al.* (1964) reported it from Bhusaval. Another species responsible for the disease is *R. stolonifer* (Ehr. ex Fr.) Lind (Pathak *et al.*, 1976).

The disease is characterised by the production of irregular water-soaked lesions. They gradually enlarge and subsequently get covered by sporangiophores which remain white when young but turn dark brown with age. The fruit finally collapses within 4 to 5 days producing watery juice.

Soft rot caused by *R. oryzae* Went was reported by Badyal (1991) from markets of Jammu. Disease is characterized by large irregular, water soaked lesions. Under humid conditions, tufts of white fluffy mycelium along with immature sporangial heads appeared on these lesions. Within two-three days whole fruit got involved becoming very soft and pulpy.

Control

Evaluation of fungicides against *Rhizopus* spp. causing soft rot of fruits was done by Sridhar (1974). The best control of *Rhizopus* spp. on grapevine and papaya was obtained by post harvest treatments with Aureofungin (100 ppm) and Dowicide (2000 ppm). Captan was also effective at 5000 ppm. Benlate and Thiabendazole were proved ineffective against this disease (Pathak *et al.*, 1976).

7.15. FUSARIAL ROTS

In our country a number of *Fusarium* spp. like *F. solani* Appel and Wr. (see 7.10, Shukla *et al.*, 1978; Pathak *et al.*, 1976) *F. fusarioides* or *F. chlamydosporium* (Gangawane and Nehemiah, 1977), *F. acuminatum* Ellis and Everh., *F. equiseti* (Corda) Sacc., *F. oxysporum* Schlecht (Lal and Arya, 1980), *F. moniliforme* Shelodon *(Arya et al.*, 1986) *F. roseum* (Tandon, 1967) and *Fusarium* spp. (Ghush *et al.*, 1964; Prasad and Verma 1976; Ratnam and Nema 1967). are reported to cause the damage to the papaya fruits.

This disease was reported from Allahabad (Arya *et al.* 1986) and markets of Jammu (Badyal 1991).

F. accuminatum. The disease starts as a water-soaked lesion and soon a cirular depressed spot develops which increases rapidly. The skin becomes soft and infected tissue pulpy. Under moist laboratory conditions white mycelial growth of the fungus develops on infected spots after 5 days.

F. equiseti. The disease was most prevalent in the months of December and January and was more common on completely ripe fruits. At first a water soaked spot appeared which increased further involving the deeper tissues also. White coloured cottony growth of the fungus was recorded on irregular older lesions. The fruit loose their shape and turn into a soft pulpy mass with in 8-10 days resulting into 100 per cent decay.

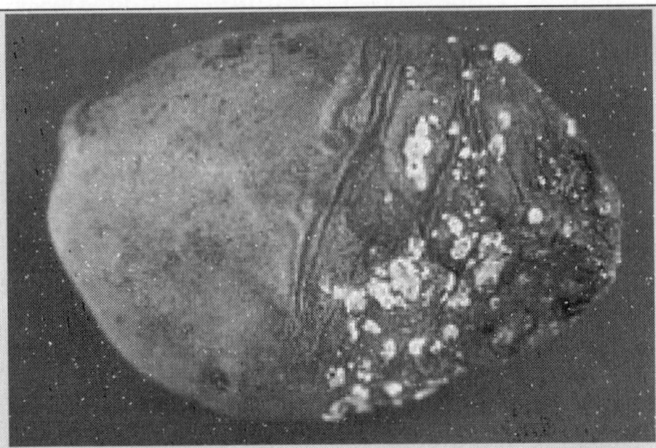

Fig. 37. Soft rot of papaya caused by *Fusarium roseum*.

F. oxysporum. When healthy fruits were artificially inoculated with the causalorganism, small circular lesions were developed which increased gradually. An extensive growth of the fungus mycelium was produced at concentric rings. The infected spot which was nearly oval in shape measured 10 × 7 cm in diameter after 7 days of incubation at 26° ± 1°C.

F. moniliforme. The disease started as a water soaked lesion and soon a circular depressed spot developed which increased rapidly. Under moist conditions, white mycelial growth of the fungus appeared on the surface after 5 days. The whole fruit rotted within 8 days and the soft pulpy tissue emitted a fermented odour.

7.16. LASIODIPLODIA ROT

In Ranchi and Bhagalpur, which are more important centres of papaya cultivation in Bihar, the maximum rot was due to *Rhizopus* sp., *Alternaria tenuis* and *Lasiodiplodia theobromae* Pat. (Prasad & Verma, 1970). Tandon (1976) obtained mummified fruits infected with *B. theobromae* from Allahabad, Kolkata, Patna and Ranchi.

Fig. 38. Soft rot of papaya fruit caused by various fusaria.

The disease manifests itself in the form of a watersoaked lesion. With the development of the disease the affected regions becomes dark brown and depressed. They are surrounded by dark green, water soaked areas. Under moist conditions the lesions get covered with the light grey fungal growth of the causal organisms and at a relatively late stage the pycnidia develop on such lesions.

7.17. PHYTOPHTHORA FRUIT ROT

Phytophthora parasitica Dastur was found responsible for disease (Tandon, 1967). It starts even when the fruits are still attached to plants. The fruit may be of any age or size. The infection starts at the stem end. With the progress of infection the fruits shrivel up and turn dark brown. In advanced stages of infection the fruits become mummified, light in weight and stony in texture.

In South China papaya epidemic occurred during April to September (Cheng and Wie, 2000). The disease occurred mainly on mature fruits. Fruit skins initially develop a light-brown colour and then become sunken and dark brown. The infected fruits rot quickly and drop easily. *Phytophthora palmivora* and *P. capsici* were identified as two causal pathogens. Spraying 250 mg/ litre Metalaxyl was used to control the epidemic.

7.18. THIELAVIOPSIS ROT

Thielaviopsis state of *Ceratocystis paradoxa* (Dade) C. Moreau was found causing soft rot of papaya fruits (Ciferri, 1927). A severe outbreak of this disease was observed in the markets of Allahabad and its suburbs in April-May, 1986 (Tewari *et al.*, 1988).

The disease starts as water soaked spots on nearly mature fruits. A black crust composed of dark coloured fungal mycelia and spores encircles the spots. In severe condition fruit may lose its texture and is deformed. The diseased fruit emit a foul odour.

Control

Five fungicides viz., bavistin, benlate, dithane M-45, tecto-40 and calixin were evaluated in laboratory by Tewari *et al.* (1988). They found bavistin and tecto-40 most effective, whereas dithane M-45 was least effective.

On the basis of *in vivo* trials they recommended use of bavistin at 1000 ppm (95 per cent disease control). Tecto-40, benlate and calixin controlled 85, 60 and 70 percent diseases as pre-inoculation treatment. Pre-inoculation treatment was more effective than post-inoculation treatment.

7.19. MACROPHOMINA ROT

A severe rotting of papaya fruit caused by *Macrophomina phaseoli* (Maubl.) Ashby, responsible for 5-20 per cent losses in Punjab during summer was reported by Kapur & Chohan (1968).

The mature fruits were more susceptible to disease, small size fruits were not conducive to good growth of the fungus although they were susceptible to infection. It was further observed that *M. phaseoli* could cause infection of papaya fruits from the uninjured surfaces under humid conditions, although the progress of the actual rotting was more in the injured than in the uninjured fruits. Thus we can say than *M. phaseoli* is esentially a wound parasite.

Control

Thiram, ferbam and captan completely inhibited the growth of *M. phaseoli in vitro.* Ziram and captan proved best protectants and ferbam was found best in post-inoculation treatment. Thiram and cuman which inhibited the growth of the pathogens *in vitro* proved less efficient on fruits. Ziram was better *in vivo* than in *in vitro* tests (Kapur and Chohan, 1974).

7.20. MEMNONIELLA ROT

It is caused by *Memnoniella echinata* (Riv.) Galloway. The disese was observed in the months of February and March in market of Allahabad (Rai *et al.*, 1982).

The disease appears as a soaked lesion, which in advanced stage becomes sunken and black in colour. The whole fruit rots within 10 days.

The fungus grows profusely in Potato Dextrose Agar medium but not on synthetic media. The fungus produces conidiophores which bear a cluster of short unbranched phialides producing dark, rough, globose, 1-celled, catenulate conidia.

Host Range

The organism attacked fruits of *Psidium guajava* L. and *Emblica officinalis* Gaertn. It failed to infect the fruits of *Achras sapota* L. and *Citrus sinensis* (L.) Osbeck.

7.21. ALTERNARIA ROT

Prasad and Verma (1970) reported occurrence of *A. tenuis* Nees from Arrah, Patna and Ranchi (Bihar). The pathogen has been earlier isolated from leaves and fruits from the plants growing in Allahabad and Varanasi, U.P. (Tandon 1967 ; Srivastava *et al.*, 1964).

On fruits, the disease is characterised by circular to semicircular patches covered over with mycelia and conidiophores. The pathogen having muriform conida belongs to Dematiaceae family of Deuteromycotina.

7.22. CHAETOMIUM ROT

Chaetomium globosum Kunze and Schm. caused soft rot symptoms which showed black pin head like perithecia (Prasad and Verma, 1970).

7.23. CURVULARIA ROT

The pathogen *C. lunata* (Wakk.) Boedijn caused petiole rot in Allahabad and fruit rot in Bhusaval, Muzaffarpur, Patna and Ranchi (Bihar) (Tandon, 1967; Prasad and Verma, 1970).

The symptoms on fruits were presence of pale olivaecous mycelia and condiophores on diseased soft rot part.

7.24. NIGROSPORA ROT

Nigrospora oryzae (Berk & Br.) Petch was recovered from fruits of papaya in the markets of Arrah, Bhagalpur, Patna and Ranchi (Prasad and Verma, 1970). The pathogen was isolated from leaves by Tandon (1976). Lesions with irregular margin were covered with mycelia and dark conidia.

7.25. TRICHOTHECIUM ROT

On fruits the disease is caused by *T. roseum* Link ex Fr. (Tandon, 1967).

Round shallow spots are produced by pathogen which later get covered by pink mouldy growth of the incitant. They cause superficial damage and do not enter the deeper tissues.

7.26. SOUR ROT

It is caused by *Geotrichum candidum* Link. ex Fr. It was obtained from Jammu markets on pressed fruits (Badyal, 1991). The injury initiated the disease. Under warm and humid conditions these areas got covered by a mass of white spores. The whole fruit turns soft and pulpy.

7.27. LEVEILLULA ROT

A new disease of *Carica papaya* was observed in Brazil caused by *Leveillula taurica* (Nogueira *et al.* 1997). The material found affected showed powdery mildew type symptoms on leaves and fruit. These were different from those caused by *idium caricae,* the well known powdery mildew pathogen. The fungus showed endophytic thallus and typical conidial morphology, which identified it as *Leveillula taurica* and its enatiomorphic form *Oidiopsis taurica.*

7.28. SOME MORE FUNGI ISOLATED FROM DIFFERENT PARTS OF PAPAYA PLANT.

Parts used for isolation	Organism	Locality	Reference
Leaf	*Phyllosticta carica-papayae Allesch*	Naini (Allahabad)	Srivastava *et al. (1964)*
	Myrothecium roridum Tode ex Fr.	Allahabad	Tandon (1967)
	Melanospora damnosa (Sacc.) Lind.	Udaipur	Tandon (1967)
Stem	*Hendersonula toruloidea Nattras*	Allahabad	Tandon *et al. (1965)*
Petiole	*Diploida papayae Theum*	Naini (Allahabad)	Srivastava *et al. (1964)*

	Colletotrichum papayae P. Henn.	Bhusaval	Srivastava *et al. (1964)*
	Phoma papayae Sohi *et al.*	Bilaspur	Sohi *et al. (1972)*
Flower	*Sordaria allahabadensis Tandon et al.*	Allahabad	Tandon, (1967)
	Choanephora cucurbitarum (Berk & Ray)	Allahabad	Tandon, (1967)
	Mucor sp.	Allahabad	Tandon (1967)
Fruit	*Helminthosporium sp.*	Arrah, Bhagalpur Muzaffarpur, Patna	Prasad & Verma (1970)
	Clasdosporium herbarum (Pers.) Link	Delhi	Tandon (1967)
	Phyllosticta carica-papayae Allesch	Naini (Allahabad)	Srivastava *et al., (1964)*
	Aspergillus sp.	Varanasi	Tandon, (1967)
	Penicillium sp. (Non-pathogenic)	Muzaffaripur	Prasad & Verma (1970)

8

Indian Gooseberry

8.1. GENERAL ACCOUNT

The Indian gooseberry, 'aonla' or 'nelli', is sometimes called *Emblic myrobalan.* Botanically it is *Phyllanthus emblica* L. or *Emblica officinalis* Gaertn., which belongs to the family Euphorbiaceae in Unisexuales of Monochlamydeae.

Its small leaves are clustered on their branchlets in such a way that each branchlet appears to be a compound leaf. Male and female flowers appear on these branchlets late in Spring.

Aonla is grown as a cash crop in U.P. It also grows wild at the foot of the Himalayas and at elevation upto 1500 metres in South India. A type called Banarasi aonla, with fruit having 5 cm diameter, is very popular in U.P., and North India, The fruit ripens in winter. Morphologically, it is a drupe with three seeds developing from a tricarpellary ovary. In South India, the main flowering takes place in July and again in February. The yeild of tree is variable. Cultivation of large fruited types, which bear a crop regularly every year, is very profitable. The star gooseberry or Ortaheite gooseberry, *P.acidus* is a related species.

The aonla fruit is one of the richest sources of vitamin C. It contains 500 to 750 mg of ascorbic acid per 100 g of pulp. This is much more than the vitamin C content of guava, tomato, or citrus fruits. Only the Barbados cherry (*Malpighia labra*) exceeds this figure. Aonla is rich in polyphenol and ascorbic acid. It is valued as an antiscorbutic, diuretic, laxative and antibiotic. One or the other part of the plant could be used in treating chronic dysentery, bronchitis, diabetes, jaundice, dispepsia and cough. The fruits are processed into "Chavanprash" and "Trifla" in Ayurvedic system of medicine. The small, round, six lobed fruits with hard pulp are astringent and not fit for eating fresh. They are extensively used for making pickles and preserves. The stringency helps to preserve Vit. C in the ripe fruit, but even then most of it is lost in the process of pickling. When the fruit is treated witn a 2 percent salt solution, most of the stringency is lost, but the major part of Vit. C is also destroyed. If the fruit is put into boiling water for a few minutes before treating with the salt solution, a much greater quantity of Vitamin C is retained. The Indian gooseberry or aonla is a potential crop for the salt affected soils. Common north Indian varieties are Banarasi, Chakaiya and Francis. The Francis is usually affected by necrosis, a serious constraint. Scientists at Narendra Dev University of Agriculture have developed varieties like, Kanchan, Krishna, Amrit and Neelum, suitable for salt affected soils.

Singh *et al.* (1996) suggested making **pickle of aonla** fruits. 1 kg good quality fruits are taken and outer skin is removed for the easy penetration of the salts. Fruits are kept in stainless steel vessel and mix with salt thoroughly. After mixing the salt (100g) keep it in sunlight for 2-3 days and turn the fruits 3-4 times daily. Add all the spices ajowan 10g, cumin 20g, coriander 100g, saunf 10g garlic 100g, chilly powder 10g and turmeric powder 20g to the jar containing aonla fruit. Mix well and keep the jar in sunlight for 3-4 days. When the pickle is ready, stir the material occasionally, add musterd oil 250g which helps improve keeping quality.

Aonla Cocktail :

Wash fully ripe aonla fruits and heat them in boiling water for 8-10 minutes to separate the fruit pulp from stone, and to remove astringency. Pulp is extracted by adding water equal to the weight of pulp and blending them in a blender. Sieve the pulp.Make syrup by boiling water and mixing sugar (1 kg). Boil it for 2-3 minutes and filter through the muslin cloth. Cool the syrup. Mix 0.75 kg syrup ginger and lime juice 200 ml. Add colour and essence if desired. Dissolve K_2OCl in a little quantity of water (1 litre) and mix with squash. Fill syrup in clean, sterilized bottles leaving 5 cm space at the neck and cork them tightly. Store the bottles in a cool and dry place and the product is ready to serve.

8.2. INSECT PESTS

8.2.1. White fly : *Trialeurodes rara Singh*

(Family-Aleyrodidae : Order-Hemiptera)

The white flies are small inconspicuous insects, which are primarily restricted to the tropical and sub-tropical countries. All species live on the leaves of plants mostly on the undersurface, and few of them are economically important.

T. rara is a common pest of *Phyllanthus emblica* L. It is a small insect with seven-segmented antennae, two ocelli and a pair of reniform compound eyes. Male genitalia with a pair of prominent claspers and in the female an acute ovipositor is present. The insects are oviparous and the eggs are stalked.

Control

By spraying (0.05%) Parathion.

8.2.2. Leaf gall : *Asphondylia phyllanthi* Felt.

(Family-Cecidomyiidae : Order-Diptera)

Cecidomyiidae are called as gall midges. These are minute delicate insects with relatively long antennae and legs. The antennae are of moniliform type adorned with conspicuous whorls of hairs. They are phytophagous and damage the plants in various ways. Majority of the species produce galls on various parts of plants. Usually leaves are affected.

8.2.3. Mealy bug : *Ferrisia virgata* Ckll.

(Family-Pseudococcidae : Order-Hemiptera)

The white-tailed mealy bug *F. virgata* is a pest of *Phyllanthus emblica*. Mealy bugs are characterised by their bodies being covered with mealy or waxy secretions. They are elongate oval in shape with distinct segmentation. Antenae are present and legs are well developed. Eggs are laid by some species and others are viviparous or ovoviviparous. If oviparous the eggs may be laid in a loose cottony wax.

8.2.4. Fruit bug : *Scutellera nobilis* F.

(Family-Scutelleridae : Order-Hemiptera)

This fruit bug is common on *Phyllanthus* sp. The insect has very large scutellum and extends to the apex of the abdomen, exposing the wings at its edge. The insect is brownish, 8-10 mm long and phytophagous.

Control

It is controlled by dusting BHC (10%) or spraying parathion.

8.2.5. Table : Some more insect pests of Indian Gooseberry

Sr. No.	Name	Family	Order
1.	*Cerciaphis emblica Patel & Kulkarni*	Aphididae	Hemiptera
2.	*Setaphis bougainvilliae T.*	Aphididae	Hemiptera
3.	*Caloptilia acidula Meyr.(Leaflet twister)*	Gracillariidae	Lepidoptera
4.	*Betousa stylophora Swinh. (Stem gall)*	Therevidae	Lepidoptera
5.	*Indarbela tetraonis M.(Bark borer)*	Metarbelidae	Lepidoptera
6.	*Virachola isocrates Fb.(Pomegranate butterfly)*	Lycaenidae	Lepidoptera
7.	*Selepa celtis M. (Hairy Caterpillar)*	Noctuidae	Lepidoptera

DISEASES CAUSED BY FUNGI

8.3. DIE-BACK OF INDIAN GOOSEBERRY

Coelomycetous fungus *Botryodiploida theobromae* Pat. is known to cause die-back diseases of various plants (Bilgrami *et al.*, 1979). Its association with the Indian gooseberry was reported by Arya *et al.*, (1987) from various orchards of Allahabad, Banda, Fatehpur, Kanpur and Pratapgarh districts of Uttar Pradesh. Rai (1982) found it causing dry rot of fruits.

A general unthrifty and stunted appearance of the tree was prominent diagnostic character of the die-back disease. Symptoms included drying of young twigs, papery bark and wood decay of the affected branches. Leaves in the diseased parts showed chlorotic appearance resembling nutritional deficiency followed by premature leaf fall. Characteristic browing of the vascular elements was observed. Generally 5-20% (in a few cases above 25%) trees were affected. Artificial inoculations in Indian gooseberry showed that within 7-10 days of inoculations, leaves became chlorotic which were subsequently shed off leaving leafless branches. The discolouration of twigs proceeded towards the main branch.

8.4. CLADOSPORIUM ROT

Two spp. of *Cladosporium* including *C. tenuissimum* Cooke and *C.cladosprioides* (Fries) de Vries were found to be associated with the fruit rot of *P.emblica.* Both the organisms produced dry rot symptoms and caused 2-5 per cent loss of the fruits (Jamaluddin, 1978).

The infection of *C. tenuissimum* was noticed from November to February. The disease started as colourless area slightly soft and subsequently progressed in a circular manner. Light brown mycelial growth of the fungus was also evident on the infected areas. The diameter of the lesions varied from 10 to 15 mm in 7 days.

At the end of season during February-March, the fruits were observed to be infected with *C.cladosporioides*. The rot showed dark brown necrotic lesions. Slight growth of the organism appeared in the necrotic cavity. The severity of infection was recorded only in mature and ripe fruits. Some freshly harvested fruits also showed this infection. One week old lesions were 7 to 12 mm in diameter. Earlier Tandon and Varma (1964) reported *Cladosporium herbarum* on fruits of 'aonla'.

8.5. PHOMOPSIS ROT OF AONLA

Lal *et al.* (1982) observed a soft rot of *Phyllanthus emblica* L. (Aonla) var. *Banarsi* in orchards as well as markets of Allahabad and Fatehpur. Isolations made from diseased fruits yielded a well sporulating culture of *Phomopsis phyllanthi* Punith. Originally, it was described on twigs of *P. reticulata* (Punithalingam, 1975).

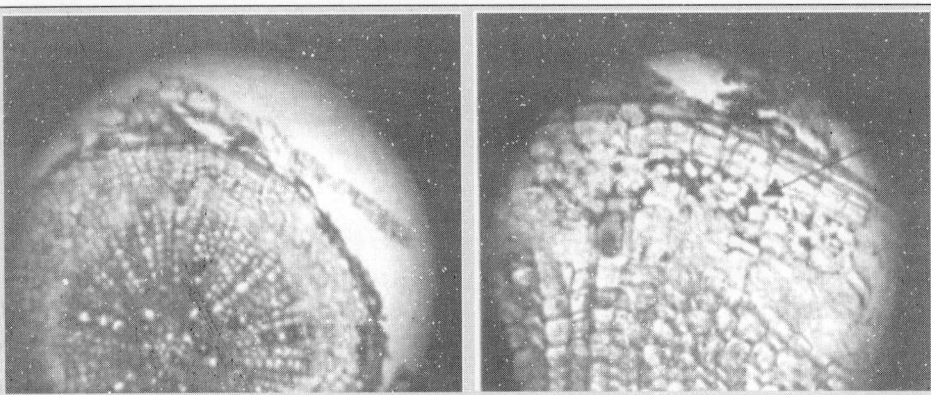

Fig. 39. Phomosis rot of Aonla.

Smoke brown to black rounded lesion developed within 2-3 days of inoculation. The diseased part later on showed olive brown discolouration with water soaked area and covered the whole fruit within 8 days. The shape of the fruit also deformed. The disease was more prevelent during Dec. to Feb. The fungus caused infection of young as well as mature fruits.

Five fungicides were tried against pathogen in three different concentrations. Use of Bavistin and Dithane M-45 at 1000 ppm and Difolatan at 1500 ppm is suggested. Blitox and Benlate could control 70% disease at the concentrations hence not recommended.

8.6. ASPERGILLUS ROT

Srivastava *et al.* (1964 *b*) reported occurrence of *Aspergillus awamori* Nakazawa and *A. niger* van Tiegh. causing surface rot of aonla.

About 8 to 21 per cent fruits of aonla rotted in storage. The susceptible host provides a congenial environment for the establishment of pathogen.

A study of metabolic pattern of the host plant revealed various changes induced during pathogenesis (Jamaluddin *et al.*, 1979). Maltose, glucose and fructose were present in 'aonla' fruits. During inocubation period, there was a gradual loss in the quantity of maltose. A synthetic oligosaccharide (Rf 0.23) appeared in diseased fruits on the 6th day and it continued to increase upto the 12th day. The formation of this oligosaccharide may be attributed to the activity of transglycosidases produced by the invading fungus. Oxalic acid was detected in the diseased fruit only. In general, the total quantity of organic acids in the infected fruits was slightly less but there was some increase in succinic acid contents of diseased fruits. Organic acids are important sources of respiratory energy in plant cell, particularly in fruits. The reduction in the total organic acid contents in the infected (aonla) fruit may be attributed to the host-pathogen interaction.

8.7 PENICILLIUM ROT OR BLUE MOLD

The occurrence of *Penicillium* sp. on fruits of *Phyllanthus emblica* L. was reported by Srivastava *et al.* (1964 *b*).

The symptoms include brown coloured depressed spot which is soft and develops blue coloured growth of fungus (due to phialides and conidia). The disease occurs more often on fallen fruits or fruits kept for storage and marketing.

8.8. PHOMA ROT

The disease was responsible for 20-25% loss of the aonla fruits in the markets of Allahabad (Srivastava *et al.*, 1986).

The disease starts as a small light brown spot on the fruits which gradually enlarges and get depressed at margin. At later stages the centre of the lesion attains a dark brown colouration. After 4-5 days minute, brownish black pycnidial bodies develop on such lesions and gradually the decay penetrates deep into the flesh. Finally, the affected tissues become brownish black and entire fruit rots within 10-12 days.

Isolations were made on potato dextrose agar medium and the causal organism was identified as *Phoma sorghina* (Sacc.) Boerma, Dorenbosch and V.Kest. Earlier occurrence of *Phoma* sp. was reported by Srivastava *et al.* (1964 *b*).

8.9. PESTALOTIA ROT

Srivastava *et al.* (1964 *b*) found occurrence of *Pestalotia cruenta* Syd. From Allahabad (U.P.).

The pathogen causes dry rot of *Emblic myrobalan*. Fruits become brown coloured and shrinked. When kept in humid conditions the fruit develop white mycelium and black coloured spores of *Pestalotia* ooze out from acervuli.

Control

The fruits, if treated with 1250 ppm Bavistin, show control of pathogen.

9

Bael

9.1. GENERAL ACCOUNT

The bael (*Aegle marmelos* Corr), belongs to the family Rutaceae, to which also belong oranges and lemons. The structure of the fruits of these two types is, however, very different. Lemon is a hespiridium and bael is an amphisarca. The bael fruit has a hard shell and the pulp contains tunnels which are filled with a mucilage.

Bael is an important indigenous fruit of India. It grows wild in sub-Himalayan tracts of the Central India, Sri Lanka, Pakistan, Bangladesh, Thailand and most of the South-East Asian countries. In India, bael tree is available in almost all the states, but Uttar Pradesh, Bihar, West Bengal, Madhya Pradesh and Orissa are main bael-growing states. It can stand swampy or alkaline soils and is propagated usually by seeds, which are sown in June. The yield per tree is 300 to 400 fruits per year. It ripens in months of March to September.

Ideal bael cultivars ought to have medium sized fruit (1 to 1.5 kg), thin rind, minimum seed content, less fibre, low mucilage with dry seed cavity, low phenolics, pleasant flavour, acceptable aroma, less prone to fruit cracking and internal fruit rotting. A few selections have been in Uttar Pradesh, Bihar and West Bengal during 1961-1971. Some popular types are Mirzapur, Darogaji, Ojha, Rampuri, Kaghji Gonda, Kaghji Etawah and Deoria Large (Srivastava *et al*. 1998).

The plant has many economic uses. The leaves are offered to Lord Shiva in prayers. Wood is used for fuel and other purposes. All the parts of the spinescent tree *Aegle armelos* are attributed with some or other medicinal properties. The plant possesses soft cork exfoliating in irregular flakes, trifoliate leaves with ovate lanceolate leaflets and greenish white flowers in short panicles. The fruits are globose, grey or yellowish with a woody rind enclosing numerous oblong, compressed seeds embedded in sacs covered with thick orange coloured sweet pulp.

Fruits have pulp 57.1%, fibre content 1.4-7.1%, acidity 0.36—0.75%, TSS 22-44% and vitamin C 15.4-18.4 mg.

The fruit contain coumarins like marmelosin (imperatorin), alleimperatorin and psoralen, β-sitosterol and an alkaloid aegein. A water soluble gum and tannic acid (more in unripe fruits) are the other constituents of the fruits (Daniel, 1991). The ripe fruits are edible and used in the preparation of cooling drinks. The fruit is used in chronic diarrhoea and dysentery and is believed to act as tonic for heart and brain. Unripe fruits are astringent, digestive and stomachic. The roots and bark are used as a remedy in melancholia, intermittent fever and palpitation of heart. Roots, containing marmin, auraptene, umbellifeone and lupeol, exhibit anti-amoebic and hypoglycaemic properties. The leaves are used as febrifuges and against ulcers, leaves contain another important alkaloid aegelinine.

Bael fruit peel constitutes about 20-25% of the total fruit weight depending upon variety (Kaushik, 2001). On an average 1000 to 1200 kg fruit peel could be obtained from one acre area. Bael peel being very hard is not relished by animals. However, in ground state it can be used with concentrate. The peel of bael fruit is recommended as a good source of animal feed for exploitation.

9.2. PHOMA ROT

A severe fruit rot of bael was observed in the orchards of Allahabad by Pandey *et al.* (1980). Isolations revealed the presence of *Phoma glomerata* (Corda) Wr. and Hochapfel.

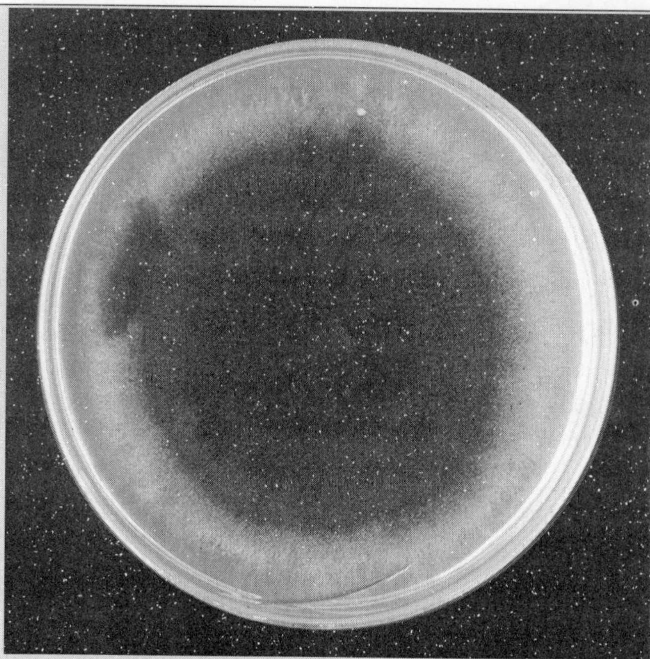

Fig. 40. Phoma glomerats rot of bael.

The fungus was highly pathogenic to immature fruit (3-4 months old) causing rot and drop when there was heavy rain or stormy wind. The infected fruit showed dark brown to black spots on the green rind. The succulent placenta and inner pericarp of infected portion become pigmented and pulpy.

When the young fruits were inoculated by injury method, typical rot developed within a week. The rot starts as water soaked lesions on the rind which turn black soon. The infected fruit later becomes soft enmeshed by white mycelial growth.

This fungus has been also isolated from infected twigs.

9.3. STALK-END ROT

Bhargava *et al.* (1977) reported immature dropping of young fruits during rainy season, due to infection caused by *Fusarium solani* (Mart.) Sacc.

The fungus appears to be highly pathogenic on immature fruits (2- 4 months old) and is observed at the stalk end of fruits. The disease is characterized by softening of rind and underlying pulp. The affected rind turns dark brown at a later stage of infection. The fruits neither shrivel nor lose their shape until pressed. Infected fruits which survive on the tree fall when there is heavy rain or strong wind.

Pathogenicity trials indicated that immature fruits are susceptible and cause fruit drop. When young fruits were sprayed with spore suspension, typical rot developed within a week through stalk ends and at a later stage the white profuse mycelial growth of the fungus entirely covered the surface of the fruits.

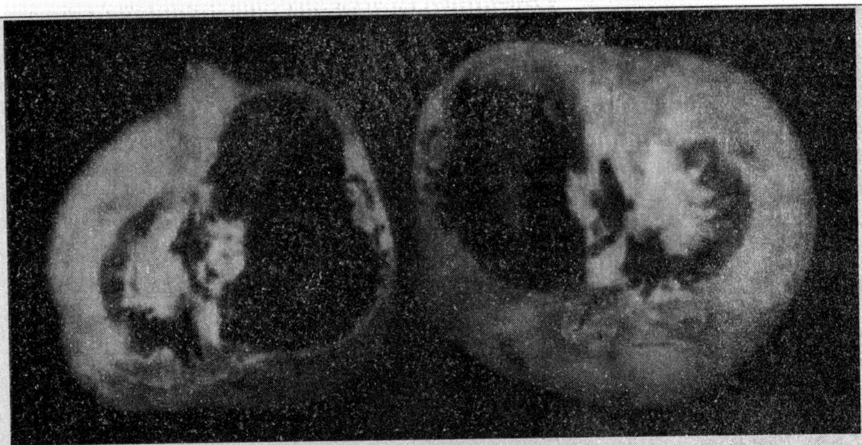

Fig. 41. Stalk end rot of bael.

Fusarium semitectum var. *majus* has been reported earlier (Mitra, 1935) on darkened young shoots of *A. marmelos.*

Control

In vitro, fungicides Benlate, Captan, Plantvax, Thiram (at 1000 ppm), Difolaton and Ferbam (at 500 ppm) were found to be effective against the pathogen, while Cercobin, Vitavax and Aureofungin (an antibiotic) were ineffective even at 2000 ppm (Bhargava *et al.*, 1977).

9.4. ASPERGILLUS ROT

It is caused by *A. awamori* Nakazawa (Arya *et al.*, 1986). During summer months, of May-June a severe post harvest rot of bael was observed. The disease resulted into 100% loss of the fruit kept in storehouse, since the infection was latent. In severe cases the white outer skin of the shell became soft and rotten inner pulp produces black sooty growth of the fungus.

The fungus showed a wide host range, it caused infection *in Malus sylvestris* L. *Pyrus communis* L., *Musa paradisiaca* L., *Psidium guajava* L. and *Citrus reticulata* Blanco etc.

9.5. FUSARIUM ROT

Like stalk end rot this is another disease caused by *F. moniliforme* Sheldon (Arya *et al.*, 1986).

During rainy season in the months of June-July, along with the black rot of bael caused by *Aspergillus* (see 9.4.), a new melady was observed. In such cases the white cottony growth of the fungal mycelium was observed just beneath the hard shell. The fungus which encircled the whole fruit turned it soft and pulpy.

Host range

When artificially inoculated, the fungus infected the healthy fruits of *Carica papaya* L., *Psidium guajava* L., *Litchi chinensis* Gaertn. and *Musa paradisiaca* L.

10

Jamun

10.1. GENERAL ACCOUNT

The 'Jamun' or Java plum, *Syzigium cumunii* (*Eugenia jambolana* Lam.) is considered to be native of India or further east. It is a common tree all over India wherever mango is found. It is not grown is regular plantations but is a common road side tree. A related wild species, *S.fruticosum*, with small, but edible fruits is grown as a wind break around orchards.

The large evergreen tree has small dark purple date-like fruits with prominent elongated seeds. The colour of the flesh is also purple and the fruit is astringent even when ripe. Some superior types have large fruits with whitish flesh. It's fruits ripen a little later in the season than mango. It is usually propagated from seeds. The fruit is usually mixed with salt before eating.

10.2. INSECT PESTS

10.2.1. *Dialeurodes eugeniae*

(Family - Aleyrodidae : Order - Hemiptera)

D.eugeniae, a fruitfly, causes characteristic punctations on both the surfaces of leaves, but more often on dorsal surface of *E.jambolana*.

It is often called aleyrodid which is a small insect with seven segmented antennae, two ocelli and a pair of reniform compound eyes. Sexual reproduction is common but parthenogenesis may occur. Males develop from unfertilised eggs whereas both the sexes develop from fertilised eggs.

Control

Control is possible by spraying Parathion (0.5%) at 15 days intervals, 3 sprays.

10.2.2. Oriental fruit moth : *Spilonata rhothia* Meyer

(Family - Eucosmidae : Order - Lepidoptera)

S.rhothia occurs on guava, jamun and mango leaves. Adults are generally small and brownish or grey with patterns on wings. Fringes of long hairs arise on the basal part of cubitus in the hindwing which is a characteristic feature of family.

Larvae web or roll up the leaves and bore into fruits.

10.2.3. Leaf miner : *Argyroploce aprobola Meyr. and A.mormopa* Meyr.

(Family - Eucosmidae : Order - Lepidoptera)

The above two insects roll up the leaves of Jamun and some other fruits like *Tamarindus* and litchi etc. The codling moth *Enarmonia* (*Carpocapsa*) *pomonella*, is one of the well known pests of apple and other fruits in the United States, belong to this family (For details see 1.2.12).

DISEASES CAUSED BY FUNGI

10.3. LASIODIPLODIA LEAF SPOT

Leaf of *Eugenia jambolana* Lam. show oval or irregular brown coloured spots. Small spots coalesce and form bigger spots. The isolation from infected leaves showed presence of a coelomycetous fungus *Botryodiplodia variispora* (Died) Z.Ambettakis. This fungus has been earlier reported by Butler and Bisby (1939) as *Diplodia variispora* Died.

10.4. CERCOSPORA LEAF SPOT

The disease occurs during rainy season. Young leaves are more susceptible. Numerous irregular vinaceous brown or deep brownish spots of variable size are characteristic of disease.

Isolations from the leaves yielded a well sporulating culture of *Cercospora eugeniae* (Rangel) Chupp. The disease was reported from Nandi Hills, Mysore (Karnataka) by Rangaswami *et al.* (1970).

10.5. FUSARIUM LEAF SPOT

Reddy (1975) reported occurrence of *Fusarium acuminatum* Ell & Ev. fromthe leaves of Jamun in Jodhpur, Rajasthan.

Symptoms include browning of oval to irregular spots, sometimes, the infected portion detach from the leaf giving a shot hole appearance.

10.6. ALTERNARIA SOFT ROT

During July to August, 1988 fruits of *Syzygium cumini* were found infected with *Alternaria alternata* (Fr.) Keis. (Tewari *et al.*, 1988).

The diseased fruits showed water-soaked lesions. These spots were pinkish in the beginning surrounded by a white margin but in advance stages more than half of the fruit rot within 8 days and then turn black with slightly raised margins. Koch's postulates were fully satisfied. Injury was found to be essential for the initial development of the disease.

10.7. ANTHRACNOSE

It is caused by *Pestalotiopsis palmarum* (Cooke) Steyaert and *Colletotrichum gloeosporioides* Penz. (Wadia and Manoharachary, 1982).

The affected fruit shows small circular to oval water soaked spots (1-3 mm in diam) on the surface. The spots coalesce and appear discoloured on the dark coloured fruit. Affected regions appear sunken on which black acervuli are formed under humid conditions.

Symptoms caused by *C. gloeosporioides* were similar to *P. palmarum*. Pink acervuli appear on sunken spot and the rotting extends to the seed.

10.8. SOFT ROT

It is caused by *Penicillium expansum* Link. ex S.F.Gray.

Irregular pale discoloured lesions develop all over the fruit surface on which the fungus sporulates profusely. The fruit becomes soft, loses its shape and emits musty odour.

10.9. BROWN ROT

Rhizopus stolonifer (Ehrenb. ex Fr.) Lind (Wadia and Manoharachary, 1982) is responsible for the disease.

Water soaked areas appear on the surface of the fruit usually near the stalk end. The disease is characterised by rapid tissue disintegration which gradually involves the entire fruit. Fruits turn soft and appear mouldy. Isolations and microscopic examination revealed presence of *R. stolonifer*.

10.10. FUSARIUM ROT

It is caused by *Fusarium semitectum* Berk. & Rav.

The infected discoloured spots (2 -5 mm in diameter) spread rapidly, turn brown, covered by a white mass of mycelium and appear powdery.

10.11. BLACK ROT

It is caused by *Aspergillus niger* van Teigh. Brown spot measuring 1-3 mm in diam appears randomly distributed on the surface of the fruit. Infected spots spread in a concentri manner and turn black with age. The fruit is covered with black mass of spores.

10.12. CURVULARIA ROT

Curvularia lunata (Wakker) Boedijn is responsible for the disease.

Circular to oval spots (2 to 4 mm in diam.) appear all over the fruit. The spots are clearly demarcated, superficial and slightly sunken. At advanced stage, the spots turn black due to profuse mycelial growth and sporulation.

10.13. TABLE : SOME MORE FUNGAL DISEASES OF EUGENIA JAMBOLANA

Part of the Plant	Disease/Pathogen	Locality	Reference
Leaf	*Neocosmopora vasinfecta Smith*	Jobner (Rj)	Goyal *et al., 1971*
Leaf	*Macrophoma mangiferae* Hingorani & Sharma	New Delhi	Waraitch, 1977
Leaf	*Pestalotia carissae Guba*	Pune (Mh)	Gopinath Nair, 1964
	P.eugeniae Thuem.	Pune (Mh)	Rao, 1963
Leaf	*Pestalotiopsis versicolor* (Speg) Stey.	Allhabad	Tandon & Bilgrami, 1961.
Leaf	*Phoma glomerata (Corda)* Wr. & Hochapfel	Jobner (Rj)	Goyal *et al., 1971.*

11

Pomegranate

11.1. GENERAL ACCOUNT

The pomegranate (*Punica granatum* Linn.) is a shrub commercially grown for its acid sweet fruits mainly used for dessert purposes, while its juice make a delicious drink. It was cultivated in the Middle East more than 5000 years ago. The pomegranate has a versatile adaptability to a wide range of climatic conditions, it is a winter-hardy, salt-hardy and drought-tolerant plant and can thrive under desert conditions.

In India, the pomegranate is a minor fruit with an area of about 800 hectares and a turnover of 40 to 50 lakhs in 1974 (Phadnis, 1974). Maharashtra accounts for more than two-thirds of the area in the country, while states like Gujarat, Uttar Pradesh, Andhra Pradesh, Karnataka and Tamil Nadu share the rest. The Bijapur district ranks first in area under pomegranate cultivation in Karnataka. Pomegranate is gaining the popularity as most important commercial fruit crop in the arid and semi arid regions of India (Koujalagi 1996). Pomegranate GKVK1, IIHR Selection G137 Alandi, Dholka, Kabul, Ganesh, Jalore Seedless, Muscut, Jyoti, Khandari, Khog, P26 and P23 have been adopted by the growers.

Jalore Seedless is propagated by cuttings as well as by air- layering. 60-70% cuttings establish if treated with 1000 ppm IBA. Last week of February to second week of March is the optimum time for planting. In arid regions, June-October is the flowering season. Fruits follow from the first week of December to the last week of February. The quality of fruit is superior, the seeds are soft and the fruit colour is brilliant red. The yield is 20-25 kg/ plant after 5 years (Bankar, 1995).

In India flowering is observed sporadically all round the year. However, there are two main seasons of flowering, namely the Spring (Ambe-Bahar) and beginning of north west monsoon (Mrig Bahar) with a smaller third one (Hasta Bahar) during the closing period of the rainy season. The crop is regulated to one season to obtain better results. Regulation of crop is done by withholding the irrigation for 2-3 months prior to flowering. Formerly there was a practice of digging the soil around the trunk and exposing part of the tree root to sun and in some cases a few roots were pruned.

Pomegranate appetizer can be made at the domestic level (Sethi and Maini, 1998). Select fully ripe fruits, take out seeds and extract juice (800 ml). Make sugar syrup by mixing water 1 *l* and sugar (1.5 kg) and add all the ingredients rose syrup (100 ml), lime juice (200 ml), table salt (2.5 g), black salt (1.5 g), ginger juice (10 ml) and black pepper powder (5 g) and mix them thoroughly.

The fruit finds wide application in both Ayurvedic and Unani traditional system of medicine. Every part of the pomegranate is used in the treatment of dysentery, diarrhoea, stomachache, dyspepsia, bronchitis and cardiac diseases. These therapeutic properties are due to the presence of betulic acid and urrolic acid and different alkaloids—pseudo-pelletierine, isopelleterine and methyl pelleterine. The unripe fruits and flowers are useful in inducing vomiting and the rind of the fruits is given in diarrhoea and dysentery. It is useful in sore throat and eyes, brain disease and chest trouble.

Dried seeds of pomegranate are termed Anardana, a valuable commodity used as condiment and for other culinary purposes. The seeds have oestrogenic activity due to the presence of 1.0 mg oestrone and 0.36 mg coumestrol (a nonsteroidal oestrogen) per 100 g of seeds (Moheam *et al.* 1988).

11.2. INSECT PESTS

11.2.1. Stem Borer

It is an important pest of pomegranate. The insect bores the stem which causes death of small twigs and ultimately loss in yield. The attack is prevented by 2-3 spraying with 2 kg carbaryl of 50% DDT in 500 litres of water at 15 days interval.

11.2.2. Fruit Borer : *Virachola isocrates*

Pomegranate butterfly causes considerable damage to pomegranate fruits. Eggs are laid on flower buds, opened flowers or young fruit. On hatching, the caterpillar bores through the fruit and feed inside it till fully grown when it comes out by making a hole in the rind for exit. The attacked fruit become worthless. The damage by the the insect is often very heavy. Bagging of fruits with butter paper gives good protection. Spraying of Deltamethrin (0.002%) and Carbaryl 50 WP (0.2%) in rotation at 21 days intervals starting from fruit-set are most cost effective measure (Pareek, 1997).

11.2.3 SHOT-HOLE BORER : *Xyleborus perforans*

Shot-hole borer *Xyleborus perforans* Wallastan (Coleptera : Scolytidae) was found damaging main trunk and branches in major pomegranate growing areas of Karnataka (Jagginavar and Naik 2001). Among 19 insecticide treatments drenching and swabbing method proved effective. Drenching with Chloropyriphos and Carbendazim (80%) and Carbaryl and Carbendazim (35.71%) proved effective.

DISEASES CAUSED BY FUNGI

11.3 POMEGRANATE WILT

The disease is caused by *Ceratocystis fimbriata* Ell & Halst.

Somasekharan and Wali (2000) reported the findings of survey conducted in Karnataka and Maharashtra during 1996-1998.

11.4 GREY MOULD

It is caused by *Botrytis cinerea* Pers.

It is one of the three major diseases of pomegranate. Decay usually starts at the calyx. As the disease progresses, the skin becomes light brown, tough and leathery. The pulp bearing seeds disintegrate into a dark mass in advanced infections. Under moist conditions, a characteristic grey mycelium appears on the affected surface (Ryall and Pentzer, 1974).

11.5 HEART ROT

It is caused by *Aspergillus niger* van Tiegh. and *Alternaria* sp. affected fruits show a slightly abnormal skin colour but internally a mass of blackened arils becomes prominent. Usually there is a black line of decay extending from the calyx to the interior part of the fruit. The disease develops while the fruit is on the tree. *Alternaria alternata* Fr. Keissler causes dark brown to black spots, turning internal tissue of fruits brown and completely rotten (Singh and Majumdar, 2000).

Control

Singh and Majumdar (2002) used Tilt, Dithane M 45, Bavistin, Topsin- M, Bayleton, Kavach in cold and hot (50°C) water as pre- and post inoculation treatment. Tilt at 0.1% and Dithane M-45 at 0.2% conc. were most effective.

11.6 ALTERNARIA ROT

It is caused by *Alternaria solani.* The heart rot is also recorded to be caused by *Alternaria solani* (Vyas and Panwar, 1976). *Alternaria* causes blackening of juicy testa of seeds.

11.7 PENICILLIUM ROT

Penicillium rot, caused by *P.expansum* and other *Penicillium reticulosum* Birkinshaw, produces water soaked spots developing blue or green spores. Infection is frequently caused if the fruits are injured. Other fungi find shelter, after its growth and eventually over grow the *Penicillium.*

Infection of pomegranate with *P.reticulosum* causes the inner parts to develop grayish green spore masses without any visible external symptoms. The possible toxicity of metabolic products of this fungus were tested on *Paramecium* and mice ; and the fungal metabolites caused the death of *Paramecium caudatum* within 3-20 min, and of mice within 3-7 days after oral administration (Osipyan and Batikyan, 1983).

12

Pineapple

12.1 GENERAL ACCOUNT

The pineapple (*Ananas comosus* (Linn.) Merrill.) is one of the most important commercial fruits of the world. It is a good tropical fruit belonging to family Bromeliaceae. It was domesticated in tropical America (Collins, 1948) and is believed to have been introduced into India as far back as 1548 A.D. *Ananas* ($2n$ = 50) and its closely related genus *Pseudoananas* $2n$ = 100) are distinguished from other Bromeliaceae by their fused syncarpous fruit. Andaman and Nicobar islands are very suitable for its cultivation. Queen and Kew varieties are grown here.

The pineapple is a xerophytic succulent, herbaceous, perennial plant. The trough-shaped leaves are spirally arranged in a dense rosette pattern. The leaves are densely covered with large trichomes covering strips of stomata located in furrows. Stomatal density is low and pores are of small size. The single spikelike inflorescence carries variable number of flowers and is terminated with a vegetative shoot or crown. Pineapple, like most succulents, fixes CO_2 via CAM pathway. Natural inflorescence development is initiated by short daylength and cool nights, but growth regulators like ethephon, 2-chloroethylphosphonic acid are used commercially to promote flowering. Pollination is not essential in self-incompatible commercial cultivars. Fruit develops parthenocarpically. By planting suckers of different sizes at different intervals and treating them with 100 ppm Ethrel (at 40- 45 leaves stage). A pineapple plant produces only one fruit during its life time. 40-50% plants come to flowering mainly during April-July (Singh, 1998).

Fully mature fruits are unsuitable for transportation. Less mature fruits are selected. Immature fruits are not shipped as they fail to develop good flavour. Care is taken to damage the crown leaves. Gibberellic acid can help in crown retention. Fruit are waxed after washing, frequently with polythylene/paraffin, carnauba/paraffin waxes. The fruits are graded on the basis of appearance characteristics. The fruit needs to be mature, firm, well formed, free of defects and with flat eyes and a minimum soluble solids content 12%. Crown size is a crucial grade component.

12.2 INSECT PESTS

12.2.1 Fruit Fly : *Daucus dorsalis* Hendel

(Family-Tephritidae : Order-Diptera)

Fruit flies are medium sized insects usually with spots or bands on wings. These insects are of immense importance as the larvae of most species are phytophagous. Variety Smooth Cayenne is not now regarded as a host for tephritid flies : Mediterranean fruit fly, *Ceratitis capitata* Wiedermann, the melon fly, *Dacus cucurbitae* Coquillet, and the Oriental fruit fly, *D. dorsalis.* Prior to 1953, pineapple fruit were fumigated with methyl bromide before importation into the continental USA. Insect disinfestation is not required for fruit imported into fly-free countries.

12.2.2. Pineapple Scale : *Diaspis bromeliea* Kerner

The pineapple scale occurs wherever pineapple is grown. Other hosts include species of *Agave, Billbergia* and *Bromelia.* It is not a major problem in Hawaiin fields. Scale can be controlled relatively easily by preharvest insecticide applications.

12.2.3. Pineapple Caterpillar : *Thecla basilides* Geyer

The pineapple caterpillars (*Thecla basilides* Geyer; *Metamasius ritchiei* Marchall ; *Batrachedra methesoni* Musch; *Paradiophorus crenatus*, Billbarg) are exotic and limited to Central and South America and the Caribbean (Rohbach, 1983). The adult oviposits on the inflorescence prior to anthesis. The larvae infest the fleshy parts of the bracts and feed inside the developing infloresence, excuding gum from the feeding chambers. Can be controlled with insecticides.

12.2.4. Pineapple fruit mite : *Steneotarsonemus ananas* Tryon

Pineapple fruit mite *Steneotarsonemus ananas* Tryon occurs universally on the growing plant, developing inflorescence, fruit and crown. The fruit mites feed on developing trichomes on the basal leaf tissue, flower bracts and sepals, causing light brown necrotic areas. Severe damage occurs when the fruits mature under drought conditions; infestation may cause death of basal crown leaves, affecting fruit quality.

12.3. BACTERIAL DISEASES

12.3.1. Fruit collapse

Pineapple fruit collapse in caused by a strain of *Erwinia chrysanthemi* which in specific to pineapple (Lim and Lowings, 1983)

Fruits normally develop symptoms at about 2 to 3 weeks before ripening ; rarely do young or ripe fruits exhibit symptoms. Infection is characterized by copious exudation of fluid from the inter fruitlet fissures, accompanied by bubbles of gas. The skin of infected fruit changes from dark purple to olivaceous green.

Bacteria is a non-sporing motile rod. On PDA yeast extract agar, whitish light yellow translucent colonies are produced.

Control

In a rotation crop, the most convenient and economical method of control is to time the forcing interval so that the opening of flowers from one forcing round does not coincide with collapsing fruits from the previous round Most farmers are tended to force at two and a half to 3 month intervals. This interval is not recommended as it allowes for maximum overlapping between open flower and collapsing fruits. Sprays on the ground with insecticide against the ant vectors have been demonstrated to reduce disease.

12.3.2. Bacterial Heart Rot

Bacterial heart rot is caused by the same pathogen that causes pineapple fruit collapse viz. *Erwinia chrysanthemi.*

Infection takes place via the stomata of the base of the young leaves. The bacteria can be transmitted by insects, rain splash and wind. The disease is not soil borne and the infection does not take place via the roots. Attempts to reproduce the symptoms by inoculation on the roots and base of the stem have not been successful.

Timely sanitation is of prime importance in bacterial heart rot control (Lim1985-86).

12.3.3. Marbled Fruit

Marbling or marbled fruit of pineapple is a bacterial disease caused by *Pseudomonas ananas* Serr. or *Erwinia ananas* Serr. The latter bacterium is now recognised as *Erwinia ananas* pv. ananas which belongs to the 'Herbicola' group. The disease in also known as fruitlet brown rot and bacterial fruitlet rot.

No external symptoms are disermble. Fruitlets at the middle of the fruit or the base, are usually more severely affected than those towards the crown. The rot can be extensive, reaching to the core, but more often it is restricted to one or more of the locules. The lesion is mottle bone brown and feels harder and drier than the surrounding healthy tissues.

12.3.4. Pink Disease

The disease is caused by a group of acetic acid bacteria that belongs to several genera, viz. *Gluconobacter, Acetobacter* and *Enterobacter* (Rohrbach and Pfeffier, 1976). Pink disease of pineapple is caused by *Poantea citrea,* a member of Enterobacteriaceae (Cha *et al.* 1977). Production of compounds in pineapple which become pink to reddish brown upon cooking the fruit pulp or juice. This colour is not induced by *E. coli, Agrobacterium tumefaciens* and others.

The symptoms are not visible externally. The exact nature of the symptoms expressed within the fruit viz. colour, intensity and aroma, depends largely on the bacterial genus that is responsible. The flesh colour can vary from light pink to brown, while the aroma, from the normal to the odour of ripe cantaloupe. The discolouration on the fruit may be entire or restricted to the fruitlets.

Control

Field experiments have shown that the disease could be reduced by dusting of Disyston 10G (disulfoton) four times on developing pineapple inflorescenes.

DISEASES CAUSED BY FUNGI

12.4. CERATOCYSTIS ROT

It is caused by fungus *Ceratocystis paradoxa* (Dade) C. Moreau. Its anamorph is *Thielaviopsis paradoxa*. It is also called black rot, black heart, water blister, soft rot or water rot. It is a universal fresh fruit problem. Fungus is a member of Fungi imperfecti. It is characterized by rotting of the fruits in storage. The fruits emit bad odour after few days. As a wound parasite the disease starts at the point of detachment of fruit. Disease tissue turns dark at later stages because of the dark coloured mycelium and conidia. On culturing, the fungus with black endo-conidia is obtained (Chandra, 1985-86).

Control

Black rot is commercially controlled in fresh fruit by minimizing bruishing of fruit during harvest and handling by refrigeration and with fungicides. Jamaluddin *et al.*(1975) recommended Benlate to be effective against Ceratocystis rot of pineapple. Bavistin, a formulation of carbendazim was equally effective for this fruit. Careful handling during picking, storage and dipping of fruits in Benomyl, TBZ or Bavistin (100 ppm) give good control (Aulakh *et al.* 1998). The stem end rot can be checked by exposure of the harvested crop to the sun for 2 hours. The cut ends of the fruit stalk should be dipped in 10% solution of benzoic acid in alcohol. According to Damayanti *et al.*(1990), gamma rays reduced the growth of *Ceratocystis paradoxa* and directly correlated with doses applied. It is also suggested to spray the packing cases and baskets with 3% formalin solution. Graham *et al.* (2000) attempted control of black heart by genetic engineering technique. Inhibition of polyphenol oxidase (PPO) was observed in genetically engineered pineapple plants.

12.5. RHIZOPUS ROT

Rhizopus stolonifer has been shown to be responsible for decay of fruits during storage, transportation and marketing. The fungus gets entry through injured portion or stalk end. The tissue becomes soft and pulpy.

12.6. FRUITLET CORE ROT

Fruitlet core rot, black spot, fruitlet brown rot and eye rot are all terms used to describe the brown to black colour of the central part of an individual fruitlet. The disease could become more important if low acid cultivars are grown. The disease is caused by a complex involving fungi *Penicillium funiculosum* Thom., *Fusarium moniliforme* Scheldon var. *subglutinas* Wollenw & Reink and the round yeast *Candida guilliermondi*. Pineapple red mite and fruit mite are associated with it. The infected fruitlet do not enlarge as rapidly as unaffected fruitlets and frequently lead to misshapen fruit (Paull, 1997).

12.7 YEASTY FERMENTATION

A large number of yeasts can attack pineapple (Paull, 1997). It is not sterile inside. In fruit with cracks yeasts start growing. This growth leads to fermentation, with bubbles of gas and juice escaping through cracks in the skin. The skin turns brown and leathery and the fruit become spongy with bright yellow flesh.

13

Nagpur Mandarin

13.1 GENERAL ACCOUNT

Citrus fruits are among the most important fruit crops in subtropical region. In 1993 world citrus production totalled about 80,000MT, which is nearly the same level as bananas and grapes. The commercially important species are sweet orange (*Citrus sinensis* (L.) Osb.), **mandarin** and tangrin (*Citrus reticulata* Blanco) lemon and lime (*Citrus limon* Burm. F. and *Citrus aurantifolia* Swing.), grapefruit (*Citrus paradisi* Macf.).

The initial citrus variety is thought to have originated in Assam region or adjacent areas in South East Asia. The cultivation of citrus trees is widespread in those tropical, subtropical and temperate regions that have suitable climate. The Vidarbha region of Maharashtra is the largest belt in India for growing the famous Nagpur Mandarin with about 0.4 million tonnes of annual production.

In Nagpur mandarin, after harvest ripening does not occur but the colour of the skin improves. Fruits with 9-12 per cent total soluble solids (TSS) and a TSS: acid ratio from 15 to 18 are considered mature and they taste good. A careful monitoring is required to test the maturity of the crop for harvest, apparently change of yellow to orange colour is the right time to test maturity standards. Dark green and immature fruits should never be harvested.

Oranges, lemons, lime and grapefruits are the principal sources of Vitamin C, folate and fiber. They are also high in a class of phytochemical called limonoids. This antioxidant has been found to be very effective against cancer. Fruit contain B carotene, xanthophyll (diol), cryptoxanthin and vialaxanthin (5, 6 epoxide) as carotenoids, which cause the orange colour of the rind and juice (Murata 1997). β carotene is precursor of vit. A and can be converted in the human body to retinol, an active vitamin compound. Some citrus fruit which are red in colour contain anthocyanin pigment.

Citrus fruit is classified as nonclimacteric fruit. Mature, sound citrus fruits show no respiratory peak and little ethylene production in normal conditions after harvest. The respiratory rate of the rind is nearly ten times as high as that of vesicles. The rind, therefore, has an important physiological role in the qualitative changes that take place during storage of the fruit. The fruits produce ethanol and acetaldehyde during storage at chilling temp. (Davis *et al.* 1974). High temp. causes shrivelling and rot by fungal organisms.

Waxing of the fruit reduces water loss, retards shriveling, but improves appearance and shine. There are two types of waxes- solvent based with acetone or ethyl acetate base and water based. The three categories of waxes depending upon their use are, storage wax, pack-out wax, and high shine wax, for giving a very high gloss on market demand. Bavistin or Benlate (1g/litre of wax emulsion) and 2,4D (500ppm in wax emulsion) can be mixed to reduce decay.

The orange quality does not improve in storage. Most of the oranges go directly to market or to processing plants. Storage is not required. Some cultivars need to degreen before marketing. Conditions of 1-5 ppm ethylene, 20-29°C temp. and 90-95% relative humidity are recommended for degreening (Wheaton and Stewart, 1973). Mandarin can be stored for 8 weeks. Ethylene induces stem and decay. Blue mold and green mold are serious diseases. Ventilation rate in the storage room influences the internal environment and incidence of decay of orange (Waks *et al.* 1985). Individually sealed packing with low density polyethylene (LDPE) film effectively reduces fungal rot.

13.2. INSECT PESTS

13.2.1. Fruit Fly : *Daucus dorsalis* Hendel

(Family—Tephritidae : Order—Diptera)

Fruit flies are medium sized insects usually with spots or bands on wings. These insects are of immense importance as the larvae of most species are phytophagous. Mediterranean fruit fly, *Ceratitis capitata* Wiedermann, the melon fly, *Dacus cucurbitae* Coquillet, and the Oriental fruit fly, *D.dorsalis* attack the fruits.

13.2.2. Oriental Psylla : *Diaphorina citri* Kaw

It is a vector of citrus greening bacteria (*Liberobacter).* It can be controlled by insecticides Monocrotophos @0.7 ml or Phosphamidon @ 0.3 ml/litre of water.Singh (1992) reported spray of 0.1% parathion. The citrus Psylla is a serious pest throughout Northern India. Its adults and orange yellow nymphs suck sap from tender shoots, leaves and even flower buds and may cause the death of shoots.

13.2.3. Citrus Rust Mite : *Tetranychus sexmaculatus* Riley

Large colonies infest leaves and fruits, causing brownish patches, and the drying of leaves and the shedding of the fruits prematurely.

Spray of lime sulphur of 0.05% wettable sulphur is recommended.

13.2.4. Citrus Red Scale : *Aonidiella aurantii* Maskell

Scale insects multiply either sexually or parthenogenetically. Most of the species lay their eggs under protective scale cover. Some of them also lay eggs in an ovisac with white cottony mass. Fecundity of the scale goes upto 2000 eggs. Incubation period is 6-9 days. Young ones are called crawlers. They move about a short time. Nymphs start feeding and become immobile in the rest of their life. Occasionally serious, infest branches, trunk and fruits, suck sap, reduce the vitality of trees.

Control

Prune and destroy affected parts. Dimethoate 0.05%, neem oil 2%, Quinalphos0.05%

13.2.5. Leaf-eating Caterpillar : *Papilio demoleus* L.

Also called lemon butterfly. Caterpillar brownish black and white when young, turning green with brown bands, feeds on leaves and defoliate trees. Control by hand pick of caterpillars. Spray 0.04% Monocrotophos or Phosphamidon.

DISEASES CAUSED BY FUNGI

13.3. FOOT ROT AND GUMMOSIS

It is caused by *Phytophthora palmivora* Butler.

In certain areas in Maharashtra, particularly around Nagpur, foot rot and gummosis is found commonly associated with die back symptoms. As a result, sudden wilting and collapse of trees has been observed in the region and finer roots of such trees are found to be rotted and fibrous. Prior to death the plant usually flowers and dies before the fruit mature. Ramakrishnan (1965) and Govinda Rao (1954) reported the disease to be common in sweet orange plantations all over South India. In Assam the fungus associated with disease is found to be *Phytophthora parasitica* (Dastur) often associated with *Fusarium lateritium* Nees (Nariani and Chaudhuri, 1985-86). Four species of *Phytophthora* can attack citrus trunk, branches, roots, leaves or fruit; *P. citrophthora, P. hibernalis, P. parasitica* and *P. syringae* All of these occur in California and most of them are found in citrus growing area of the world (Klotz,1973).

13.4. STEM END ROTS

It is caused by *Lasiodiplodia theobromae* Pat., *Alternaria citri, Phomopsis citri*and *Colletotrichum gloeosporioides* Penz.

The *Lasiodiploida (Botryodiplodia)* stem end rot is an important disease of Nagpur mandarin causing around 21-26% loss. The spores of the fungus are produced in pycnidia develop on dead twigs of the plant. The spores are dispersed by the rain splash and germinate beneath and around the calyces. Infection may take place any time from the fruit set to harvest.

Stem end rot caused by *Alternaria citri* usually develops in cold storage of the fruits. Infected fruit show browning at stem end because the fungus grows internally along the central core of the fruit. Infection of *Alternaria* also takes place in the orchard during fruit development but remains quiescent till harvest. Air borne conidia of the pathogen settle around the button and remain latent till harvest. The disease is important during commercial storage of Nagpur Mandarin. Since the fungus is resistant to all the fungicides, it can only be controlled by keeping the button of the fruit green if it is harvested by clipping, by the use of 100-500 ppm 2,4 D. Alternaria brown spot of citrus is also reported from Spain (Vicent *et al.*, 2000).

13.5. ANTHRACNOSE

It is caused by *Colletotrichum gloeosporioides* Penz. and *Glomerella cingulata.*

It is common problem in Nagpur Mandarin. The conidia of the pathogen are produced abundantly on dead twigs of the plant and are dispersed by the rain and settle on fruit surface where they germinate and form a quiescent structure called appressoria. These appressoria germinate and form intercellular dormant hyphae up to 2-3 cell depth of the rind. The latent infection becomes active after fruit harvest during storage. Ge Hong Xin *et al.* (2000) reported reasons for the rampant occurrence of citrus anthracnose in South China.

13.6. SOUR-ROT

It is caused by *Geotrichum candidum* Link.

An important post harvest disease of Nagpur mandarin is caused by a wound pathogen *Geotrichum.* The fungus invades through injuries and bruising and grows in the form of filamentous yeast like growth on the fruit surface. The infected portion becomes soft and pulpy.

Control

G. candidum is resistant to Benomyl and Bavistin. Guazatine and potassium sorbate show some control of this fungus but Guazatine is not registered in India. The feasible control of the disease is careful handling of fruits to minimize the injuries to fruit surface or up to some extent refrigerated storage.

13.7. PENICILLIUM ROT

It is caused by *Penicillium digitatum* Sacc. and *P. italicum* Wehmer.

Green mould and blue mould rots are important post-harvest diseases of citrus fruit. Two species of *Penicillium* attack fruit on the tree or in packing house, transit, storage and market. Green mould and blue mould are wound parasites, these spread by contact. Green mould is much more prevalent than blue mould. Blue mould is more common in long term storage at low temperature. Green mould is an efficient ethylene producer. Both rots first become apparent to casual examination as soft, watery, decolorized spots in the rind. Soon after the mould begins to appear on the surface. Respiratory rate and fungal rot of the fruits are suppressed by low temp. Minimum temperature requried for the growth of *Penicillium* sp. is 6-8°C.

Control

The disease can be reduced by avoiding the injury to the fruit. Fruits which get infection by *Aspergillus* and *Penicillium* should be washed with Chlorine water (1000 ppm) chlorine conc. Washed fruits should be treated with Bavistin, Benomyl 1 g/litre water. Such fungicidal treatment reduces 70- 80% disease (Naqvi, 1996). Smilanick and Sorenson (2001) suggested control of post harvest decay of citrus by calcium polysulfide. According to them, green mould was reduced 80% by immersion in liquid lime sulphur (30-40°C) for 1-4 min.

13.8. ASPERGILLUS ROT

Aspergillus niger van Tiegh. invades through wounds and injuries, and causes soft rot of the fruit. The infected area of the fruit turns pale yellow with a characteristic halo leaving the shrink intact cuticle on the macerated peel. In advanced stage infected surface is covered with black mass of conidia which disseminate to contaminate the whole lot.

13.9. GREY MOULD ROT

It is caused by *Botrytis cinerea* Persoon.

13.10. DISEASES CAUSED BY BACTERIA

13.10.1. Citrus Canker

Citrus canker causes extensive damage to citrus industry. The damage is comparatively more on acid lime and less on mandarin and sweet orange. The diseased plants are characterized by the occurrence of conspicuous raised necrotic lesions that develop on leaves, twigs and fruits.

Xanthomonas citri causes pin-point spots that attain a diameter of 3-10 mm. In leaves, a yellowish halo often surrounds the lesion. Severe infection results in defoliation, die back and premature fruit drop. Canker causes fruit losses from abscission and non-marketable quality of fruits. The bacteria survives in lesions in leaves, twigs and branches which constitute the source of inoculum to spread the disease from season to season. Bacteria from the canker lesions are disseminated mostly by wind splashed rains. The long-distance dissemination takes place through diseased planting material.

Control

Canker has become an endemic disease in our country. Effective control can be achieved by continuous care and attention paid by the grower. Canker incidence can be reduced considerably by using canker-free nursery stock, pruning and destruction of infected twigs, 3-4 sprays of copper oxychloride (0.3%) or Bordeaux mixture (1%) and Streptocyeline (100 ppm) at monthly intervals after the onset of monsoon (Das and Singh, 2001). Rangaswami (1957) found streptomycin sulphate effective. Patel and Padhya (1964) observed 3 sprays of sodium arsenite and copper sulphate effective against the disease .

* Spraying of neem cake solution (1kg/20 litres of water).
* Pruned twigs should be burnt immediately.
* New flush of the tree should be sprayed properly.
* After pruning, the tools should be disinfected with 1.5% sodium hypochlorite solution.

13.10.2. Citrus Greening

It is caused by *Liberobacter asiaticum* Greening is reported from different citrus growing region of India. Sweet orange and mandarin are more susceptible to this disease. It is common in citrus orchards of Marathwada and Vidarbha regions (Das and Singh 2001). Its incidence was recorded on 1-3% plants. Trees of older age and poor health have higher incidence of the disease.

Greening infected leaves are small, upright and frequently have Zinc deficiency like symptoms with green veins and chlorotic interveinal areas. Diseased leaves also show various types of chlorotic mottling in severe cases. Leaves become almost chlorotic with scattered dark green islands. Symptoms are often seen on a part of the canopy.

Greening is caused by bacteria living in photosynthetic tissue. By molecular biology technique it is characterized as *Liberobacter asiaticum*. It is transmitted by infected bud woods and oriental Psylla vector (*Diaphorina citri*).

Control

The control involves removal of unproductive trees and their replacement by disease free budded plants.

The disease-free certified plants should be raised. To control vector Psylla, suitable insecticides like monocrotophos @ 0.7 ml or Phosphmidon @ 0.3 ml/litre of water should be sprayed to control its spread. Although tetracyclines have been reported to suppress greening symptoms.

13.11. DISEASE CAUSED BY PHYTOPLASMA

During surveys conducted in 1998-1999 in Nagpur Mandarin orchards of Nagpur district, an unusual type of disease with multiple sprouting was noticed. Incidence of the disease was recorded (Ghosh and Singh, 2002). Pathogen culture was established in screen house by graft inoculation on one year old Nagpur mandarin seedlings after taking samples from field infected plants.

Phytoplasma infected Nagpur mandarin plants produce two distinct types of symptoms. In certain cases, entire plant shows symptoms of thin, linear shoots with multiple sprouting and unusual flexibility. Affected plants produce smaller, slightly thickened chlorotic leaves and produce no flowers or fruits. This type of disease symptom was earlier reported as cirtus rubbery wood disease (Ahlawat and Chenulu, 1985). Phytoplasma culture has been established in screen house on Nagpur mandarin. However, till date no symptom has been recorded on etrog citron, acid lime, mosambi, rough lemon and trifoliate orange seedlings (Ghosh and Singh, 2002). Witches broom disease of lime (*Citrus aurantifolia*) (WBDL) is caused by *Phytoplasma aurantifolia.*

13.12. DISEASES CAUSED BY VIRUSES

13.12.1. Tristeza

The name 'Tristeza' was first suggested by Moreira (1942) to describe the appearance of the diseased citrus trees. Biological and serological characterization of citrus Tristeza virus (TV) has been done for the isolates obtained from Morocco (Zemzami *et al.,* 1999).

Failure and eventually death of plants occur. It was first reported from South Africa (1899) then from Java, Argentina, Brazil etc. Existence of tristeza in Indian orchards was suggested by Lal (1953) thereafter, Vasudeva and Capoor (1958), Reddy and Govinda Rao (1961) and Nariani *et al* (1965) reported the presence of Tristeza from different parts of India. Between 1950-1960 lakhs of trees were killed in India owing to root rot and die back which were attributed to Tristeza disease. Usually, affected trees show a relatively gradual decline, and linger on for years before they finally wilt.

Few trees show wilting symptoms overnight and completely dry up in 2-3 days. Hence Tristeza is also called a quick decline disease (Reddy and Murti, 1990). Flexuous thread-like rods about 10 to 12μm × 2000μm (microns) are present in phloem of affected trees (Bar-Joseph *et al.* 1970) The disease is transmitted by *Toxoptera citricida* and *Mazus persicae.*

Control

Since Tristeza virus and its most efficient vectors are present all over India, the control of the disease in rather difficult. Eradication and use of resistant rot stocks are common preventive methods.

13.12.2. Seedling-Yellows

This malady of citrus seedlings is caused by unknown virus (Fraser, 1952). According to Wallace (1965), there are two different viruses causing Tristeza and Seedling yellows. Wallace and Drake (1972) concluded that seedling- yellows was caused by a mixture of two or more components, one of which might be tristeza. Fraser (1968) has said that no one has studied seedling-yellows separately, possibly because it could not exist alone. This disease is transmitted by budding and vectors as listed above.

13.12.3. Citrus mosaic

It was reported from Andhra Pradesh by Reddy *et al.* (1972) The disease is present to an extent of 5.84% of orchards (Murti, 1983). Affected leaves show irregular yellow or light green patches, alternating with normal green areas. It resulted reduction of leaf length and fruits developed depressed yellow patches.

13.12.4. Yellow-corky Vein Disease

The disease occurs on Sathgudi on a limited scale especially in Kurnool district of Andhra Pradesh. As this is perpetuated through grassing, it is necessary that the parent trees should be free of this disease. It is safe to collect budwood from mature nucellar sweet orange trees.

13.12.5. Psorosis

Psorosis can be of two types Psorosis-A, Psorosis-B, Blind pocket, concave gum, crinkly leaf and infectious variegation (Fawcett, 1939; Wallace 1959). However, the infections variegation and crinkly leaf viruses are mechanically transmitted and so considered as strains unrelated to psorosis group (Wallace 1968b).

The most common feature of psorosis A is the scaling of the bark on the trunk or large limb. Affected areas are almost circular. In the beginning they are like smaller blisters, eventually coalescing into large scaly patches girdling trunks and limbs.

Psorosis is transmitted through grafting. Seed transmission is also reported (Childs and Johnson, 1966). No insect vector is known.

Control

The best control of this disease is through use of virus free bud wood and using a disease free parent.

13.12.6. Exocortis Disease

This disease was reported originally by Fawcett and Klotz (1948). Exocortis (*Exo*-outside and *cortis* -pertaining to the bark) is characterized by dying of narrow strips of outer bark, tending to separate from the inner live-bark and slowly peeling off as it withers. These symptoms are seen below the bud union on the root stock portion of susceptible stocks.

The disease has been reported from India by Nariani *et al.* (1968), Patil and Worke (1968) and Kapur *et al.* (1973). The disease is transmitted by grafting. Exocortis free budwood should only be used.

14

Phalsa

14.1. GENERAL ACCOUNT

The phalsa (*Grewia asiatica* L.) is an important catch crop in the commercial fruit orcharding. It is mostly used as fresh fruit. Fruits are used for making juice and squash, which is tasty and nourishing in hot summer months. It can be grown in wide range of soils. Being a hardy plant it can withstand drought conditions and can be grown under adverse soils and climatic conditions. It can tolerate alkalinity as well. Yet its another asset is that it is comparatively free from pests and diseases possibly due to the severe pruning treatment given to it. Due to these qualities, the growers exploit not only its wide adaptability but also its uses. Botanically *G. asiatica* or *G. subiinaequalis* DC belongs to family Tiliaceae. It bears small yellow flowers in axillary cymes. G. *asiatica* is native to India in the Baroda area of Gujarat.

The fruits are reddish violet in colour, round few mm in size. These contain 55 to 65% juice with citric acid 2.8%, sucrose 11.7% and traces of vit C. Ripe fruits are sub-acidic and good source of vitamin A and C (Chand *et al.*, 1985-86). The pectin content is rather low. The fruit can be converted into bottled beverages, an excellent ready to serve drink during the hot weather when its fruiting coincides. Due to its sub-acid taste, it is equally relished in its fresh condition.

Its cultivation is favoured around big cities, where fruit find a ready and quick sale an important determining factor for its cultivation because of the poor keeping quality of its fruit. To a certain extent, it is grown in a cultivated form in Punjab, Uttar Pradesh, Andhra Pradesh and Maharashtra. Otherwise, it is chiefly grown as an inter crop in other fruit trees. When grown as pure crop vegetables can be grown with it. Pruning is generally done in mid-winter in the North (Dec.–Jan.). Yield of 6-10 kg per plant is quite common.

Hayes (1960) was of the opinion that phalsa responds to manuring. Nijjar (1969) reported that application of 1 kg N/plant as farmyard manure or CAN or their combination gave high yield. Singh (1967) found that regular manuring gives better yield. It is reported that the plants receiving deficient supply of N, P or K made acute stunted growth with little branching and few smaller leaves, the effect being most pronounced due to N deficiency and the N, P, K deficient plants failed to produce flowers. Singh *et al.* (1981) reported that application of 0.4% Zinc sulphate at pre bloom stage and after berry set improved the juice content. Application of 0.4% ferrous sulphate alone or in combination with Zinc improved the berry size (Mitra, 1988).

14.2. INSECT PEST

14.2.1. Bark Eating Caterpillar

The bark eating caterpillar is the only serious pest which attacks the plant severely (Singh, 1967).

Control

The control measures consist in destroying the live caterpillars and plugging the shelter tunnels with kerosene, petrol, carbon- bisulphide mixture of ethylene dichloride and carbon tetrachloride (3:1) or mixture of creosote and chloroform (1:2). Spraying the trees with 0.25% DDT or BHC is also effective

DISEASES CAUSED BY FUNGI

14.3. ALTERNARIA LEAF SPOT

Causal organism : *Alternaria* state of *Pleospora infectoria* Fuckel.

The disease was reported for the first time from Hissar by Gupta and Madaan (1979). The disease is common in occurrence, wherever, phalsa is cultivated. Symptoms of the disease appears on the leaves in the form of brown to dark brown, irregular spots starting from the margin of the leaves. Later on spots fuse to form big blighted patches. Concentric rings are also present on these spots and severely affected leaves get defoliated. This disease is caused by *Alternaria* state of *Pleospora infectoria* Fuckel.

The mycelium of the fungus is septate, irregularly branched and light grey in colour. Condiophores become light brown, septate, measuring 27.2-68.0 $\times$ 3.4-6.8 μm. (microns) Conidia are formed in chain, light brown, muriform with small beak. Both longitudinal and transverse septa are present, 1-6 cross septa, measure 13.6-54.4 $\times$ 6.8-13.6 μm.

Control

This fungus survives in soil that becomes the primary source of infection. The secondary infection takes place through the spores present in air. The disease can be controlled by spraying Dithane Z-78 (0.2%).Two sprays should be given at two weeks interval.

14.4. RUST OF PHALSA

Causal organism : *Dasturella grewiae* (Pat. & Har.) Thirum.

Symptoms : The disease appears in form of yellow to light brown colour pustules having uredospores. In severe cases fall of leaves take place. The dwarf varieties are less susceptible as compared to tall ones.

Control

The disease can be controlled by spraying Dithane M-45 or Dithane Z-78 (0.2% conc.). Two or three sprays should be given at 2 weeks interval (Gupta and Madan 1985).

14.5 CERCOSPORA LEAF SPOT

Causal organism : *Cercospora grewiae* Srivas. & Mehta.

This disease was first reported by Srivastava and Mehta in 1951 from Kanpur, U.P.

Symptoms. The disease was noticed in the rainy season of July- Aug in 1951. Small lesions developed on both dorsal and ventral side of leaves. The lesions are covered with white crust of the fungal growth. Later on patches enlarge in diameter having blackish spot in the centre. In due course the black growth aquires cushion like appearance. Such erumpent, cushion like spots are numerous all over the leaves and size of the spots become 0.5 mm in diameter (Chand *et al.* 1985/1986).

The most notable feature of this sp. is the extra matrical growth all over the leaf area, which consists of masses of conidia, conidiophores and superficial mycelium. The fungus mostly survives in diseased debris which become the primary source of infection. The secondary infection takes place through spores (conidia are present in air).

Control

The disease could be controlled by spraying Dithane M-45, 0.2%. Two or three sprays should be given after two weeks interval.

14.6. PHYLLOSTICTA LEAF SPOT

Causal organism : *Phyllosticta grewiae* Diet. apud Syd. & Butler.

The disease was first reported by Chona and Munjal (1955) from Delhi. Occasionally the disease spreads in severe form and causes considerable damage to fruit crop.

The symptoms are present in form of brown coloured spots, which become circular measuring 3.5-9.0 mm in diam. (mostly 6-8 mm) amphigenous scattered. The pathogen survives in diseased debris. The disease is caused by *Phyllosticta grewiae* Diet. apud Syd. & Butler.

14.7. DISEASE CAUSED BY PHYTOPLASMA

14.7.1. Phytoplasma Disease

A bushy stunt disease of phalsa was caused by phytoplasma and was for the first time reported from India by Rishi and Ray Chaudhary (1981). Internodes are shortened, leaves and flowers are very small in size. As a result a number of very small leaves and flowers accumulate together to form a cluster. Either there is no fruit setting on these twigs or they remain very tiny and do not ripe. The suppression of symptoms by tetracycline and observation of mycoplasma like bodies within the sieve tube cells has been observed.

15

Litchi

15.1. GENERAL ACCOUNT

The litchi [*Litchi chinensis* (Gaertn.) Sonn.] is the most important sub-tropical evergreen tree and belongs to the family Sapindaceae. Litchi fruit is famous for its excellent quality, characteristic pleasant flavour and for attractive red colour. Litchi is a native of South China. According to Blume, Cochin, China and the Philippines islands are the lands of origin of litchi (Popenoe, 1920). From China litchi reached Eastern India first via Burma. India ranks second in the world from production point of view. With the increasing popularity of exotic fruits on the worldwide market, litchi production has steadily grown in importance. World litchi production is currently about 35,000 tons most of which comes from China, India and Taiwan (Underhill *et al.* 1997). It is commonly grown in Muzaffarpur and Darbhanga in Jharkhand, Gorakhpur in U.P., Dehradun in Uttranchal, Kangra in H.P., Gurdaspur in Panjab, Araku valley of Andhra Pradesh and Hoogly in West Bengal.

Litchi belons to subfamily Napheleae. The common name of this fruit is variously spelled litchi, lichee, lychee, leechee, lichi, laichi, lici and so on. Yule *et al.* (1903) stated that the pronounciation in north China is lee-chee while in southern part of the country, it is ly-chee. In English it is pronounced as litchi or lychee.

Litchi tree grows to a height of 6-9 m with spreading branches and a dense light green shining foliage. The leaf is compound, mostly paripinnate with three pairs of leaflets. Leaves arranged alternately, exstipulate, petiolate, petioles about 4 cm long, leaflets 2-9 opposite or nearly so. Litchi can grow in a variety of soil particularly in fairly deep, well drained loam rich in organic matter. According to Groff (1921), the litchi seems to do best on dykes of low land where its roots can always secure all the water needed, and where they are even subjected to periods of immersion. In some places it grows on high land but not so successfully. Humidity is another factor for the successful cultivation of litchi. The dry hot winds in summer months during the fruit development are very harmful.

Litchi is liked very much as a table fruit all over the world. In China it is very popular in dried or canned state. Dried litchi is known as litchi nut, which tastes somewhat like the raisin. Litchi fruit consists of about 60% juice, 8% rag, 19% seed and 13% skin. Litchi fruit are considerably rich in sugar, the range of sugar content in the fruit is 6.7% to 18% (Singh and Singh 1954). 70% of the sugar is reducing sugar. Litchi is also an excellent source of vit. C ranging from 40.2 to 90 mg/100 g but it contains insignificant amount of protein, fat, pectin and minerals specially calcium, phosphorus and iron (0.7%).

China, being the home of litchi, possesses the largest named and described varieties. Wu Ying Kuei reported 74 varieties. Sun Yit Hoong is the earliest variety in China but tends to have large seeds and a thin flesh. Noh Mai Chee has the reputation of being the best Chinese variety (Stephens 1955). Common Indian varieties include, Early seedless, Muzaffarpur, Rose scented, Bambai, Kolkata, Dehradun, Gulabi etc.

Like other fruit litchi is also raised by two methods, namely (*i*) sexual—by seed, and (*ii*) asexual—by vegetative means like air layering, pot layering, stem cutting etc. The young litchi plant is very delicate and if proper care is not taken at the time of planting, the mortality after planting is heavy. Planting is not advisable when the weather is either too dry or too wet. Training of young litchi plants to establish a good framework is necessary. Once the desired shape and a strong framework is achieved, no pruning is usually necessary.

There is no general consensus with regard to type of growth promoter found best for accelerating the quality of litchi, but it has been established that GA_3 increases acid content in the fruit. Veera and Das (1973) reported best results with 2,4-D 40 $mg\,l^{-1}$, while Khan *et al.* (1976) recommended NAA 20 $mg\,l^{-1}$ and 2,4,5-T 10 $mg\,l^{-1}$. Foliar sprays of zinc (4.12$\,l^{-1}$) and 2,4,5-T (10 $mg\,l^{-1}$) to increase the fruit size and quality are recommended Litchi is a nonclimctric fruit and as such does not ripen once harvested.

15.2. INSECT PESTS

15.2.1. Red Rust : *Aceria litchi Keifer*

Erinose mite causes red rust in Himachal Pradesh at an elevation of 600 m. Small deep excavated pits on the ventral surface of new leaves are initial symptoms. These pits are lined by thick velvety, dull green growth, the erineum, the colour of which changes to chocolate-brown in a few days. The leaves curl up apically or double over vertically forming hollow cylinders. In case of epidemic, the mite infest flowers and even epicarp of the newly formed fruits. As a result the fruit set and yield are reduced drastically. The infestation coincides with the appearance of the new flushes.

Control

Pruning of affected twigs and spraying of Dicofol (0.05 per cent) or Dimethoate (0.05%)

at monthly intervals helps to minimize this malady up to 50 per cent (Sharma *et al.* 1996). Wettable sulphur @ 5 g/litre can also be used to control it.

15.2.2. Litchi Fruit Borer

This is a serious problem in litchi fruits. The larva enters the young fruit near the pedicel and makes pin hole like structure. The entrance hole often heals and the fruit appears healthy. A very small pile of saw dust, however, remain stickingnear the entrance hole. The larva feeds around nut and damages 15-80 per cent of the edible fleshy part of the fruit. The infested fruits start falling before harvest (Sharma *et al.* 1996).

Control

Spraying Endosulfan (0.05%) or Monocrotophos (0.036%) from fruit set to 20 days before harvest at monthly intervals reduces its infestation up to 50 per cent.

15.2.3. The litchi stem borer : *Aristobia testuda* Voet

(Family-Cerambycidae, Order-Coleoptera)

For the first time observed in China. Litchi stem is made hollow by this insect. It also attacks guava and other fruits.

15.3. LITCHI FRUIT DISORDERS

Litchi suffers numerous physiological disorders.

15.3.1. Fruit Cracking

Fruit cracking is a serious problem, particularly in the low elevations, reducing yield up to 30-35 per cent in cultivar Dehradun. Apart from varietal differences, sharp fluctuations in the day and night temperature coupled with heavy irrigation after dry spells contribute towards fruit cracking. The brownish blotches develop in the skin of fruits and later on the skin cracks. The affected fruit rot and become unfit for marketing.

The application of 20 ppm NAA or 1.0% $ZnSO_4$ at fortnightly intervals from pea stage to fruit maturity in combination with irrigation of trees at 20 per cent depletion of available soil moisture helps reduce the intensity of fruit cracking up to 60-70 per cent. Suitable tree species to be used as windbreaks for litchi orchards should be selected.

15.3.2. Zn Deficiency

Zinc deficiency has emerged as a burning problem in some of the litchi orchards. Initially symptoms appear on terminal flushes. In severe infection all the flushes are affected. The affected leaves become bronze in colour, narrow, stiff and cup backward, giving a cup shaped appearance. In severe deficiency conditions, there is almost total stoppage of flushing and the death of twigs and even branches. About 50-60 per cent of the trees are suffering from the deficiency.

Control

Foliar spraying of $ZnSO_4$ (0.5%) on new growth flushes helps overcome this problem.

15.3.3. Pericarp Browning

The main post-harvest disorder associated with litchi is pericarp browning. It can be caused by a wide range of different stress, such as climatic conditions prior to fruit maturation (Sharma *et al.* 1996), Disease (Huang and Scott, 1985), Desiccation (Scott *et al.* 1982) and heat injury (Wong *et al.* 1991). Litchi pericarp browning is caused by rapid degradation of phenols by the fact that peroxidase POD, a similar oxidative enzyme, is also present in the litchi pericarp. High activity of POD is reported during fruit storage. Recently Jiang and Chen (1995) recorded increased POD activity during pericarp browning, further supporting the involvement of POD.

Control

Sulphur dioxide has been used in South Africa, Thailand and Israel as an alternative treatment to control physiological browning. The growing awareness of the need to replace the chemical treatments such as SO_2 with more sustainable treatments has seen increased interest in physical methods such as heat. Kaiser (1995) reported that immersion of fruit in hot water 98°C for 30 s followed by a low pH treatment.

DISEASES CAUSED BY FUNGI

15.4. ASPERGILLUS ROT

Causal organisms—*Aspergillus niger* van Tiegh.

A. flavus Link ex Fr., *A. variecolor, A. nidulans* (Eidam) Winter, *A. quadrillineatus* Thom and Raper.

The disease usually appeared first at the stalk end of the fruits as light brown lesions encircling the stalks, which later on changed into dark brown in colour. Black conidial heads appear after 5-6 days. Some white colour deposition is also seen in the infected regions. Causal organism is *A. niger*.

In certain cases disease starts in form of circular, brownish areas which enlarged with age and became depressed in centre. The central portion turned grey and finally purplish in colour. Olive green conidial heads appear. Causal organism—*A. quadrillineatus* Thom & Raper.

15.5. LASIODIPLODIA ROT

Causal organism—*Lasiodiplodia (Botryodiplodia) theobromae* Pat.

The disease appeared as water soaked brownish discoloration. It extended irregularly on either side but more often towards the stalk end of the fruit. Rotting of the pulp usually was faster than rotting of the rind. Under humid conditions light grey cottony mycelia were observed over the infected tissue, minute pycnidia developed later on.

15.6. CYLINDROCARPON ROT

Causal organism—*Cylindrocarpon tonkinens* Bugn.

Large water soaked lesions appeared on the surface of fruit. Initially, disease symptoms were perceptible on injured portion of fruits. As the disease advanced, the decayed area became depressed. Affected fruit emit an odour of fermentation.

15.7. COLLETOTRICHUM ROT

Causal organism—*Colletotrichum gloeosporioides* Penz.

The first lesions produced by the pathogens were small, circular, slightly depressed and water soaked. Later on they coalesced to form large, irregular spots. The acervuli were provided with dark brown multicellular setae.

15.8. PESTALOTIA ROT

Causal organism—*Pestalotia* sp.

Lesions sually appeared first as light brown discoloration, which became russet-coloured after 5 to 6 days. As the spots enlarged, their central zones became slightly depressed, and later on minute black acervuli with viscid spore masses appeared either singly or in cluster on the infected region. Other sp. of *Pestalotia* reported is *P. pauciseta* Sacc.

15.9. FUSARIUM ROT

Causal organism—*Fusarium semitectum* Berk. and Rav.

In the early stage of infection the diseased fruits showed pinkish dicolouration with a border surrounding the infected regions. Gradually the diseased portion was covered completely with white mycelia which later on turned pinkish in colour. The pulp of the fruits also became light pink which on examination revealed to contain numerous conidia.

15.10. ALTERNARIA ROT

Causal organism—*Alternaria alternata* (Fr.) Keissler.

Light brown circular spot were observed in the beginning, which changed to dark brown along with olive green central part. In the advanced stages of infection the affected area of the fruit was covered by mycelia and spores. The rotted portion also presented soft and water soaked appearance.

15.11. SOME MORE FUNGI REPORTED WITH POST-HARVEST ROT OF LITCHI

Fungus	Reference
Agrostalagmus sp.	Roth 1963
Curvularia sp.	Roth 1963, Coates *et al.* 1994
Dothiorella sp.	Coates *et al.* 1994
Geotrichum candidum	Tandon and Tandon 1975
Monilia sp.	Roth 1963
Mucor sp.	Roth 1963
Nigrospora sp.	Roth 1963
Penicillium sp.	Prasad and Bilgrami 1973, Scott *et al.*1982
Phoma sp.	Coates *et al.* 1994
Phomopsis sp.	Scott *et al.* 1982
Rhizopus sp.	Roth 1963
Stemphylium sp.	Roth 1963
Trichoderma sp.	Scott *et al.* 1982

15.12. CONTROL OF FUNGAL DISEASES

Strategies for the control of litchi post-harvest disease include the use of post-harvest fungicide, refrigeration, heat treatment, packing shed and orchard hygiene, correct packaging materials and pre-harvest insect control (Lonsdale 1988). A range of fungicides has been evaluated for diseases control in Litchi including Benomyl, Prochloraz and Imazalil. Prasad and Bilgrami (1973) employed 23 fungicides Zerlate, Cupravit, Blitox, Fytolan, Ferbam, Borax, Agrosan as post harvest treatment. They have recommended coating of SOPP (0.25%) treated fruit with wax emulsion just after harvesting in order to get best results.

SO_2 which is used in some countries to prevent pericarp browning in litchies, has also been shown to have fungicidal action (Swarts, 1989). Heat treatment and biological control are some other methods.

16

Post-Harvest Disease Management

16.1. GENERAL ACCOUNT

India is one of the leading producers of fruits and vegetables in the world but unfortunately, average Indians do not get the basic daily requirement of fruits and egetables, making our Human Development Index (HDI) very low. Rapid increase in population in India has resulted in food scarcity, nutritional deficiency and environmental insecurity. Appropriate post-harvest management cannot only provide adequate food and nutrition but also play a significant role towards environmental security. Attention to the concept of post-harvest food loss reduction as a significant means to increase food availability, was drawn by the World Food Conference held in Rome 25 years ago. The UN passed a resolution calling for a 50% reduction of post-harvest losses. India is one of the horticulturally rich country of the world but unfortunately, unlike other horiculturally rich countries, average Indians do not get the basic daily requirement of fruits and vegetables (Roy 2001).

In U.S.A. the losses for fruits and vegetales amout to 14% of the harvesyed crop (USDA, ARS, 1965). In India the bulk storage facilities are not adequate. The post-harvest losses may be as high as 40 or even 50% or more, considering the cumulative losses at various points such as transit, apex fruit and vegetable bulk sale points or mandis, subsequent fruit markets, storage godown, retail shops and consumer place. According to Roy (2001), production of fruits and vegetables during 1998-1999 as per the National Horticultural Board was over 131 million tonnes. At least 50% of the total production of fruits and vegetables in the country is lost due to wastage and value destruction, and the cost of this wastage is estimated to be ₹ 23,000 crores each year. Taking the loss rate at 50% the net availability of fruits and vegetables should be about 66 million tonnes. As per the specifications of the National Institute of Nutrition at least 300 g of fruits and vegetables are to be consumed by an individual per day for a balanced diet.

Vidarbha region of Mahrashtra is famous for Nagpur Mandarin (*Citrus reticulata* Blanco) in India. Nagpur oranges are famous for quality and taste. About 67000 ha area is under orange cultivation in the central India with an annual production of 0.3-0.4 million tonnes. The serious neglect in postharvest management of Nagpur Mandarin results in very high losses of this valued fruit. The survey on its post harvest losses from farm level to New Delhi market reveals 20.91% losses in market packed fruits when transported by truck and 23.27% by train. In farmhouse packed fruits, however, the losses were marginally less (18.17%) by truck and 20.53% by train (Naqvi 1996). The predisposing factors for post-harvest diseases are ; high orchard inoculum due to poor plant-protection management, almost negligible postharvest fungicidal treatments before packing and marketing, rough and multiple handling of fruits and use of unventilated packing boxes and carriers.

Fruits and vegetables are living beings and they carry out transpiration, respiration, ripening and other biological activities, which adversely affect the quality . In addition, because of their high moisture content fruits are inherently liable to deteriorate, especially under tropical conditions, and finally become unmarketable. Fruits and vegetables can be preserved in the fresh form by increasing their shelf-life or can be processed into different products. The shelf-life can be increased by storing them in a low temperature room or in controlled atmosphere storage in properly designed packages or wrapping etc.

Some of the significant developments in the area of post-harvest management are, use of corrugated fibre board boxes, pre treatment, pre cooling and development of cool chain and cold storage of fruits and vegetables. The working group on Agricultural and horticultural crops, has suggested various measures particularly for reducing losses of perishables and making handling more cost effective (Ghosh 1997). Some of the important changes occurring in harvested produce are listed in Table 1.

Table 1. Changes in harvested produce

Changes	Process	Examples and Significance
Water loss	Transpiration Evaporation	Unattractive appearance,, texture changes,, weight loss shrivelling.
Carbohydrate conversion	Enzymatic	Starch to sugar, beneficial in bananas and pear Detrimental in sweet corn
Flavour	Enzymatic	Beneficial in bananas and pear, detrimental in others
Softening	Pectic enzymes, Water loss	Detrimental
Colour	Pigment synthesis or destruction	Detrimental or beneficial
Decay and Rot	Pathological, physiological	Detrimental

16.1 CARE IN HARVESTING

The post harvest diseases in orange are caused by *Penicillium*, *Aspergillus* and *Geotrichum.* It can effectively be reduced by simply avoiding the injuries to the fruit during harvest, handling and transport. The fruits should be harvested in dry weather and the nails of harvesting person should be well trimmed to avoid injuries to the fruit surface.

Manual harvesting, a common practice in out country, is highly time consuming and labour intensive operation. It should be replaced by creatively designed, user friendly machinery and implements and needs to be popularised with an innovative package of training, education and improved product orientation. Mechanical harvester is a must but it should be kept in mind that the plants especially of the fruit crops must not be damaged.

16.2 CARE IN HANDLING

The handling of harvested product is of utmost importance. Any lapse in case of any stage determines the amount of injury sustained. For fragile fruits of strawberries it is preferable to harvest them directly into plastic baskets (until it reaches the consumer). In unhygienic conditions the wounds are likely to become infected. It was observed that failure to renew chlorinated washing water results in presence of heavy inoculum of fungal and bacterial pathogens as chlorine in the form of sodium hypochlorite rapidly loses its disinfectant properties.

According to Naqvi (1996), the harvested fruits should be collected in plastic crates or the container having smooth walls. They should be transported in well cushioned, tightly stacked carriers to the packing house. Sanitation of the packing houses also reduces the chances of fruit contamination and further decay.

16.3 STORAGE AT LOW TEMPERATURE

The importance of cool temperature for the storage of fruits was well known to the people in ancient times. Caves and pits provided areas that were used for preservation of food as they were much cooler than outside temperature. For best results the storage temperature must be kept constant, since the optimum temperature say as little as 5°C can hasten ripening and prove fatal to the product.

The lowest limits of storage temperature must be determined for each item. Low temperature stress is undesirable for fruits such as apples and pears. The degree of injury varies with the product. Some commodities are severely damaged by even slight freezing, whereas others less susceptible may undergo a number of successive freezes without perceptible permanent injury. Storage temperatures are kept above the point at which commodity freezes.

Based on the principle of evaporative cooling, a low cost easily adaptable zero energy cool chamber has been developed at IARI, New Delhi. The chamber is made up of brick and bamboo frame. Chamber is thoroughly soaked with water, maintains a temperature of 15-17°C. The chamber is useful for short storage of fruits (Roy 2001).

16.4 ORGANIC VOLATILES AND RIPENING

Numerous organic volatiles *i.e.* acids, alcohols, esters, aldehyde and ketone are produced by ripening fruits. Ethylene an organic volatile is produced during the ripening process and accounts for two-thirds of the total carbon lost in volatile form. External application of ethylene influences the respiration and ripening processes. Thus it has been commercially used to accelerate uniform ripening of mangoes and bananas. In some instances *i.e.* degreening of oranges is done by ethylene gas as it causes destruction of chlorophyll associated with the peel.

Fruits can be divided into climacteric and non-climacteric types. In apples, pears, bananas there is a sudden surge in respiration rate known as climateric rise, while on the other hand fruits as citrus, grapes, cherries and pomegranate the respiration rate remains constant. These fruits do not have starch reserves hence must be plucked after fully mature and ripe.

Bananas, pears, avacados are some of the fruits, which prefer to their optimum when off the plant. When ripened on plants bananas become mealy, lack flavour and are subject to splitting and subsequent decay. They are best ripened at 18°C with 90-95% relative humidity while the fruit is coloured. Ripening can be hastened by holding initial temperatures at not more than 21°C for the first 18-24h and at 19°C thereafter. Prolonged high temperature increases deterioration and decay.

16.5 POST-HARVEST CHEMICAL TREATMENTS

Harvested oranges should be washed with chlorine water (1000 ppm chlorine concentration) and then thoroughly rinsed with fresh water to remove traces of chlorine. Washed fruits should be treated with Bavistin, Benomyl (1g/*l* water) by dipping them for 5 minutes, if it is done manually (Naqvi 1996). Such post harvest fungicidal treatement controls 70-80% post harvest diseases caused by *Penicillium* and *Aspergillus* up to 3 weeks. The fungicide can be mixed with wax. The wax coating of fruit not only reduces physiological weight loss but also provides extra shine to the fruit surface. The above post harvest operations can quickly be done on mechanical packing line where washing, cleaning, fungicide + wax treatement, drying the wax and grading can precisely be done.

Response of ethylene on degreening of Nagpur orange was observed. Fruit firmness declined and TSS content increased while acidity decreased on exposure of fruits to ethylene. Mechanical wax coating (Stay fresh high shine wax at 2.5% solids + Carbendazim a systemic fungicide at 2000 ppm) improved shine as compared to non-waxed fruits.

Control of post harvest browning of litchi fruit by Sodium bisulphite and hydrochloric acid is suggested by Ming *et al.* (1997). Litchi fruits were soaked in 1% $NaHSO_3$ containing 0.1,0.5 and 1% HCl for 2-30 min. Sulfite residues were mainly restricted to the non-edible skin with a residual level of 4.2 ppm in pulp after 6 days of storage. It is concluded that treatment of fruit with $NaHSO_3$ and HCl is an alternative method of preserving red colour in litchi.

16.6 PROPER PACKAGING

Wood on account of its favourable properties has remained the main packaging material for fruits and vegetables. However, to maintain ecological balance, there is an urgency to identify un-wanted and unsuitable usage of timber. Packaging has been identified to be one of the most important areas where such substitution is not only possible but also obviously desirable. Corrugated fibre board containers consume one third of the wood required for timber boxes of the same size. These boxes can also be fabricated from craft paper made from bamboo, long grasses and many other types of agricultural residues like bagasse, paddy, cotton silk, jute stick and wheat straw and recycled paper and card-board. Ventilated partition bearing CFB are developed by IARI.

Bruishing of fruits is found to be maximum in the wooden crates with straw and paper and minimum in the fruits packed in CFB boxes with partition. The cushioning materials used for packaging fruits and vegetables in woden boxes are dry grass, paddy straw, leaves, paper shreds etc. All these materials end up as garbage and add to environmental pollution in cities. On the other hand CFB boxes use either moulded trays or cardboard partitions which can be easily recycled.

Corrugated fibre-Board boxes are hygienic and light in weight. The boxes can be turned water resistant by the use of a suitable adhesive, or coating with wax or a plastic film. They can also be made stronger by reinforcing the board or box. A box constructed with a combination of plywood dand corrugated fibre board gives a high stack-load capacity. In multi-trip packaging, a recent introduction is the box made of corrugated polypropylene board. The boxes are hygienic, sturdy and have a high bursting strength. They are light in weight and are water resistant.

17

Bio-Control of Fruit Rots and Application of Botanical Pesticides

17.1 GENERAL ACCOUNT

Chemicals are used to control pre- and post-harvest decay of fruits (Eckert and Ogawa, 1985, 1988). These may give undesirable odour and residues may be toxic to consumers. Many of these fungicides are found to be mutagenic and carcinogenic (Norman, 1988). Fruit rot pathogens have even developed tolerance to major fungicides (Spotts and Cervantes 1988). Fruits and vegetables have especially been placed under scrutiny because of being contaminated with potentially carcinogenic compounds such as 'Alar' (Chalutz and Wilson, 1990). In some markets, sale of fruits with detectable residues of post-harvest fungicides are even prohibited (Willett *et al.* 1992). Therefore, all these issues have stimulated the need to search some alternative sources for controlling post-harvest rots of fruits and biocontrol (Wilson and Pusey, 1985 ; Janisiewicz, 1990) as well as application of biopesticides (Chiu 1984, Lal and Srivastava, 2002) have emerged as latest technologies.

17.2. BIO-CONTROL : A NOVEL APPROACH

Bio-control is the reduction of inoculum density or disease producing activities of a pathogen in its active or dormant state, by one or more microorganisms, accomplished naturally or through manipulation of the environment, host or antagonist, or by mass introduction of one or more antagonists. This method of control is very complex and numerous interactions are known to affect its efficacy including the fruit cultivar maturity, wound severity, pathogen antagonist, ratio and environmental conditions (Janisiweicz and Yourman, 1989.).

One of the earliest observation of potential biocontrol agent was reported by Gutter and Littauer (1983), who isolated a strain of *Bacillus subtilis* from citrus fruit that inhibited the growth in culture of 10 citrus fruit pathogens. *B.subtilis* also controlled brown rot of peach caused by *Monilinia fructicola*. Yeast controlled blue mold caused by *Penicillium expansum* and grey mold caused by *Botrytis cinerea* on pome fruits. A saprophytic strain of *Pseudomonas syringae* also provide complete control of blue and grey mold of pear. *Trichoderma harzianum* controlled anthracnose (*Colletotrichum musae*) of banana (Kanapathipillai, 1988) and grey mould (*B.cinerea*) of strawberries.

17.3. MANAGING ALIEN OR ARTIFICIALLY INTRODUCED ANTAGONISTS

It is an exceptionally productive area to explore because of the following considerations :

1. Greater control of the environmental conditions in the post-harvest phase.
2. Direction of biocontrol agents towards effective site.
3. More cost effective.

A number of post-harvest diseases of various fruits and vegetables have been controlled by introducing a antagonist but the fact remains that most antagonist have been tested against a particular pathogen and that too in respect of a particular crop, therefore, the practical utility for control of a number of post harvest diseases of a particular fruit poses a big question mark. There are, however, limited antagonists which are reported to be effective against major rots of citrus and apple (Wisniewski *et al.* 1988) offered by yeast *Debaryomyces hansenii*.

The fungus *Trichoderma viride* has also been observed to control apple rots due to *Alternaria alternata, Botrytis cinerea* and *Rhizopus stolonifer* (Palazon and Palazon, 1988). Similarly another species of *Trichoderma*, viz., *T. harzianum* has also been reported to be effective in checking storage rots of carrots caused by a number of fungi (Transomo 1989). However, such fungi with broad spectrum activity are fewer.

Chalutz *et al.* (1989) have summarized the information on the mode of action of antagonist that have been used in the post-harvest environment. Antibiotic production, however, appears to be a major mode of action of many of the antagonists identified so far. In addition to antibiotic production, nutrient competition, direct parasitism and possibly induced resistance do appear other modes of action.

17.4. PARASITE-ANTAGONIST INTERACTION

Injuries pave way for the pathogen or fruits may have latent infection. To control wound invading pathogens, effective antagonist must be able to colonize wounds and suppress or inhibit the pathogen. Dipping the fruit in suspension of microbial culture would tremendously affect the population and consequently resistance to pathogen.

17.5. THE WOUND AS THE SITE FOR INFECTION AND BIO-CONTROL

In harvested fruits micro-ecology of the wound takes on special significance. Most rot organisms cannot invade host tissue directly but must infect through wounds inflicted primarily during harvesting and processing. The wound site is an important battle ground for the pathogen and the host. If antagonistic microbes are involved, the wound becomes an important site of their interaction with pathogens. Infection and subsequent rot development at a wound site depends on the interaction of the host's defenses, the pathogen and the antagonists. A bacterium *Bacillus subtilis* is an effective antagonist against brown rot of peaches. It produces an antibiotic iturin, which is capable of protecting wounds in the absence of the bacterium. When wounds of peach are treated with *B. subtilis* or iturin, and subsequently inoculated with spores of *Monilinia fructicola*, spore germination of the pathogen is inhibited at the wound surface (Hazen, 1989). A complex interaction involving the antagonist, pathogen and the host defence system has been found with the non-antibiotic producing east antagonist *Pichia guillermondii* (Wilson, 1989).

It is also found that the extra cellular matrix produced by the antagonistic yeast imparts some protection to wounds, suggesting that the antagonist may induce a resistance response in the host. In addition, *P.guilliermondii* attaches directly to the walls of pathogens such as *Botrytis cinerea* and produce enzymes, which can partially digest its walls (Wisniewski *et al.*, 1991).

17.6. INDUCED RESISTANCE BY MICROORGANISMS

Induced resistance is being recognized as an important and manipulable form of resistance in vegetative plant tissue (Kuc, 1987). Similar mechanisms of resistance are probably operable in harvested commodities.

Part of the mode of action of epiphytic antagonists to plant pathogens may be through the induction of resistance response in the host plant. Swinburne (1973) found that unripe Bramley seedling apples produced benzoic acid in response to invasion by *Nectria galligenia.* He determined that protease's from the pathogen would elicit this response. Evidence has been found that part of the mode of action of antagonists against certain citrus fruit rotters may be through induced resistance. The use of microorganism to induce resistance in harvested commodities warrants exploration. Induction may be possible through use of non-biological elicitors or from elicitors derived from pathogens or antagonists.

In considering microbe induced resistance in post-harvest commodities, some questions are unanswered. For instance, do pre-harvest infection by microorganism induce resistance response in fruits and vegetables to post-harvest infections ? Also, is resistance imparted to a portion of a fruit or vegetable transmissible to other portions ? Induced resistance in vegetative cucumber and watermelon plants has been transmitted to the fruit (Caruso & Kuc, 1979).

17.7 PROSPECTS OF COMMERCIALIZATION OF BIOCONTROL AGENT

U.S. National Academy of Sciences report says that 60% of the oncogenic risk 196 *Bio-Control of Fruit Rots & Application of Botanical..* from pesticides comes from fungicides. Many of the fungicides are withdrawn from market due to safety reasons. Benomyl and Botran were the major fungicides used on peaches to control *Monilinia* and *Rhizopus* rots respectively are no longer available for these purposes.

A critical need exists for alternative to synthetic fungicides for the control of post-harvest diseases of fruits. The development of biological control (living fungicides) is already at advanced stages of commercialization.

It is important that chosen antagonists are genetically stable and non-fastidious in their nutrient requirements. In order for it to be economically feasible to grow large quantities of an antagonist it must grow well on cheap agricultural by-products such as cotton seed meal. (Hofstein *et al.* 1991). Also, selected antagonists must be 'tough' and able to survive adverse environmental conditions. They must be amenable to powdered or flowable formulations having adequate shelf life.

An ideal antagonist would be one which is an effective as chemical means of control and that too at low concentration, control many pathogens, economical and safe or in other words lacks toxicity to man and animal, able to survive in adverse conditions. Among potential antagonists yeast has appeared to possess most of the attributes enumerated above (Janisiewicz, 1988).

Bacillus subtilis (Strain B-3) for the control of stone fruit rots was the first organism patented as a post harvest biocontrol agent in USA (US Patent No.4, 764, 371). Bio-fungicide MTR-35 obtained from *Trichoderma harzianum* and Glio Gard from *Gliocladium virens* are effective against soil borne plant pathogens. While selecting an antagonist there are certain other considerations. It should not be related to pathogens. Otherwise it may pose more problems. The metabolic or toxis chemical produced by antagonist should not be hazardous to consumer in any way, we hope friendly biofungicides will capture the market soon.

17.8. BOTANICAL PESTICIDES IN DISEASE CONTROL

The distinct propensity towards a trek back to nature has become evident in the recent past, especially in the field of pesticides. This change is not sudden, but is the result of sheer necessity that has been realised out of indiscriminate use of synthetic pesticides leading to threat to people's health, the environmental pollution, resistance to these chemicals and consequent pest resurgence. Marshelling regenerative resources such as plants and their products, for newer strategies of pest management is the demand of the century. The millenia old interaction and co-evolution between pathogens and plants makes the latter ideal sources of new products for use in ingenious strategies in the management of pest control.

Search for environmentally sound pesticide received an impetus following the publication of **Silent Spring** (Carson 1962). Fawcett and Spencer (1970) emphasized the use of naturally occurring compounds obtained from higher plants in controlling plant diseases in view of their low phytotoxicity. Pradhan *et al* (1952) reported the feeding deterrent property of neem seed kernel suspension against desert locust, *Schistocera gregaria*. Subsequently, several bioactive ingredients were isolated from various parts of the neem tree, more notable being the isolation of meliantriol and Azadirachtin. The botanicals that have shown potential in pest control during the past few years are neem, Chinaberry, West Indian Mahagony, custard apple, African marigold, French marigold and Thunder god vine (Lal and Srivastava, 2002). Srivastava and Lal (1997) reported inhibition of spore germination of *Alternaria alternata* causing soft rot of pear and pomegranate by leaf extract of *Ocimum basilicum*. Leaf extract of *Lantana camara* inhibited spore germination of *Curvularia tuberculata* causing soft rot of pear. Inhibitory effect of *Lantana* extract was recorded for *Penicillium citrinum, Aspergillus ustus* and *Chaetomium cymbeforme* (Shah and Arya, 2000). Neem leaf extract checked completely fruit rot of cucurbitacious plants (Krishna *et al.* 1986). Shatre *et al.* (1996) found neem, garlic, ginger, onion and tulsi effective against Xanthomonads.

Water soluble fractions of *Aegle marmalos* Corr., *Eucalyptus occidentalis* Skeels, *Lawsonia inermis* L. *Strichnos nux-vomica* l. and *Ephedra foliata* Wallich were tried in 25 and 50% concentration to control Phomopsis fruit rots of grapes and guava. On the basis of their effectiveness *Eucalyptus, Aegle* and *Strychnos* were selected for *in vivo* studies. Use of 50% crude extract of *Aegle* and *Eucalyptus* is recommended (Arya 1988A). Percentage spore inhibition of *Phomopsis psidii* and *P.viticola* in three different concentration of neem, *Aegle marmalos, Crataeva religiosa, Ephedra foliata, Eucalyptus occidentalis, Lawsonia inermis, Ocimum sanctum* and *Strichnos nux vomica* was observed. After *in vivo* trials application of Eucalyptus extract is recommended (Arya 1988B). Post inoculation treatment was effective in guava while pre-inoculation treatment was better in case of grapes. Although *Strichnos* extract was more effective due to presence of toxic alkaloids it can not be recommended for safe use. Arya and Mathew (1990) reported control of *Phomopsis* and *Aspergillus* rot of chiku by leaf extract. *Orobanchae, Calotropis* and *Cuscuta* gave 50% spore inhibition in *Phomopsis* sp. and extracts of *Orobanchae, Eucalyptus* and *Cuscuta* were effective against *A.niger* and gave 52.9, 43.3 and 41.1% inhibition respectively.

Smearing the fruits with oils have also been tried. Six fixed oils were tested against fruit rot pathogens of grapes and guava, mustard, linseed and castor were selected for *in vivo* use. Use of linseed oil is recommended. Smearing of neem oil resulted in bitter flavour of grapes is not found suitable although it reduced the disease incidence. There is a need to test the refined oils also which do not have pungency of sulphur compounds (Arya 1988A). Use of Cashew nut oil, *Eucalyptus* oil and neem oil is suggested for control of four fungi (Shah and Arya 2000).

17.9. MODE OF ACTION

Studies on effetiveness of plant extracts revealed that the toxic activity depends on the nature and amount of the active principal. Kovacs (1964) illustrated that onion juice inhibited the development of swarm spore of *Phytophthora infestans* at its 1:100 dilutions. The

Anagalis arvensis inhibited the growth of *Colletotrichum papayae* at 1:200000 dilution (Nene and Thapliyal, 1969). Garlic and *Bignonia* may be effective due to presence of sulphur compound Allicin in them. *Eucalyptus* leaves may be effective due to the presence of volatile compound Cineole in them (Arya 1988 B). The effectiveness of fungitoxic extract depends on plant part used, the stage at which harvested, used as fresh or dehydrated form, and the nature and amount of active principal present.

All plants contain medicinal compounds, which are toxins and antitoxins, oxidants and antioxidants, nutrients and antinutrients depending on their action or reaction on the pest. The degree of their biocide character decides whether it is a potential phytobiocide or not.

17.10. ATTRIBUTES FOR A GOOD BOTANICAL PESTICIDE

* They should have low or no toxicity to higher vertebrates and nonpathogens.
* Extracts should be effective at low concentration.
* It should be easy to grow, harvest and formulate.
* It should not be a rare or expensive plant.
* The pesticides should be sufficiently effective to control a broad range of target pests.
* It must have sufficient residual action without being too persistent.
* They should possess more than one way of action, to reduce the chances of pest getting resistant to the pesticides.
* Plant species preferably a perennial, and not gets destroyed when the pest control material is harvested from it.

17.11. FUTURE PROSPECTS

Plant based pesticides have attracted the attention of biologists since last few decades and a large number of plants are screened for their phytobiocide properties. The apprehension that large scale use of plant based fungicide/insecticide may lead to resistance has not been sustained. These crude extracts contain several compounds, which act on several processes. These are safer to use as they are from plant origin so less harmful to human system. They are biodegradable in nature. Efforts should be made to prepare crude or purified aqueous or organic formulations and made them commercially available to the users. India has vast resources. A large number of plants are available and should be exploited for their potential use as biocide. It is befitting that our country takes a lead in promulgating and practising this non-violent approach to pest control.

18

Fruit Rots and Biochemical Changes

18.1 GENERAL ACCOUNT

Nutritional value of fruits chiefly depends on their quality and concentration of sugars, vitamins and other essential substances. Various fungi and bacteria not only blemish, disfigure or cause rot to a number of fruits but the post-infectional biochemical changes reduce their food and market values considerably. Numerous studies conducted during the last few decades have established the fact that both ualitative and quantitative changes occur in infected fruits. Fungi are very selective in their nutritional requirements. They influence the stored substances by absorbing them or by converting some of the complex forms into simpler ones. Some of them may be utilized directly and other may need specific enzymes before they can be used.

A perusal of literature on the changes incurred during pathogenesis in various fruits clearly revealed that the quantity of various free and bound amino acids and organic acids was altered and a gradual decrease in sugar and Vitamin C content was observed in the advancement of disease (Bhargava and Arya, 1983).

Realising the importance of post-harvest diseases, Stevens and Stevens (1952) mentioned, ''Of all losses caused by plant diseases those occur after harvest are he most costly, whether measured in monetary terms or in man hours''.

In the present chapter work done on the subject has been reviewed. Earlier, Tandon (1967, 1970) and Bhargava (1977) have given some useful information on this aspect, while dealing with some post-harvest diseases of fruits and vegetables. Changes caused by different fungi on some important fruits during pathogenesis, are discussed below :

18.2 CHANGES IN SUGARS

Fruits are generally rich in sugars/carbohydrates. Their value greatly depends on the quality and concentrations of these sugars. In general, with the maturity of fruits the amount of the sugars usually increases. The pathogen which are responsible for spoilage of fruits in storage and transit grow mainly at the expense of various carbohydrates present in fruits. When infection is established in the host, further spread of the pathogens or the disease will depend upon the availability of nutrients to the pathogen at the infection site.

During pathogenesis, usually a loss of complex carbohydrates occurs which is due to the production of extracellular enzymes by pathogens. There is a simultaneous increase in the quantity of simple sugar such as glucose and fructose, etc. Later on, the quantitities of these sugars also decrease with an incubation or storage period.

As early as in 1962, Tandon and Bhargava reported that sucrose and maltose present in healthy tissues of pineapple fruit were not traceable after seven days in infected (*Botryodiplodia ananassae*) tissues, instead only glucose and fructose could be detected in the diseased tissues.

Sucrose, glucose and fructose are known to be present in healthy fruits of guava, papaya, sapota and banana. Ghosh *et al.* (1964) observed a marked difference in comparative utilization of these sugars by different pathogens, *Pestalotia psidii* and *Phoma psidii* hydrolysed the entire sucrose of guava fruit within six days while *Gloeosporium psidii* and *Diplodia natalensis* were very slow hydrolysers. All the organisms except *D. natalensis* consumed glucose earlier than fructose from the infected tissues. In case of papaya fruit, *Gloeosporium papayae* also caused rapid breakdown of sucrose. The rate of hydrolysis of sucrose was faster than that of *Colletotrichum papayae* and *Fusarium* sp., both isolated from same host.

Considerable loss of sucrose content was observed in orange fruit infected with *Curvularia lunata* (Singh and Tandon, 1970). Similarly *Cladosporium oxysporum* and *Drechslera rostrata* infecting loquat and capegooseberry, respectively utilized their total sugar contents within 5 days (Singh, 1980). The response of these two pathogens was, however, different towards glucose and fructose, the other two sugars present in the two fruits. The former utilized fructose earlier than glucose while latter consumed both these sugars simultaneously in 15 days. Rai (1982) working with some fruit rot diseases observed complete hydrolysis of maltose and sucrose within 4 days in sapota and Indian plum (Ber) infected with *Phomopsis sapotae* and *Pestalotiopsis palmarum*, respectively. *Cheatosphaeropsis truncata* var *indica*, unlike these two fungi failed to hydrolyse sucrose completely in banana fruit even after 12 days of incubation. The above three pathogens along with *Periconia saraswatipurensis* infecting grape berries depleted both these hydrolytic products (glucose and fructose) simultaneously from their respective hosts but the time taken by *P. palmarum* was lesser (8 days) than others (12 days).

Phomopsis psidii infecting guava (Lal *et al.*, 1985) hydrolysed sucrose completely within 5 days, while *P. viticola* failed to hydrolyse it completely in the diseased grape berries into 15 days of incubation (Arya and Lal, 1986). Preferential utilization of fructose was observed in both the cases by these two fungi. *In vitro* studies dealing with these pathogens (Arya, 1982) revealed that when they were grown in equi-mixture of glucose and fructose (5 g/l each) substituted in modified Asthana and Hawker's medium 'A', both the sugars were utilized simultaneously after 4 and 5 days by *P. viticola* and *P. psidii*, respectively. Thus, hydrolysis of complex sugars and preferential utilization of any monosaccharide in different fruit by a pathogen can be directly correlated with the quality and quantity of various carbohydrates and other nutritional substances present in the host fruit. The delayed hydrolysis of sucrose *in vivo* by *P. viticola* may be attributed to the high sucrose concentration and the presence of an additional oligosaccharide (Rf 0.45) in the grape berries.

Results obtained by Chandra and Tandon (1963) for 'Bombay green' and 'Rasthali' varieties of banana infected with *Fusarium oxysporum* f. *cubense* indicate loss of sucrose in infected fruits without any change in the quantity of glucose and fructose. Traces of four unknown sugars Rf 0.05, 0.11, 0.28 and 0.18 were, however, observed in diseased fruits on the 24th day. According to Tandon (1967), the behaviour of *Pestalotia sapotae* on sapota was slightly different from that of *B. theobromae* on the same host. The quantity of sucrose was greatly reduced but both glucose and fructose increased on the 4th day. No sucrose was left on the 12th day when only traces of glucose and fructose were available.

Starch in culinary and table varieties of banana could not be traced out after 15 days in fruits infected with *B. theobromae*, while no appreciable change occurred in control fruits (Williamson and Tandon, 1965).

Chaudhary *et al.* (1980) reported considerable decrease in total sugars and an increase in reducing sugars of both healthy and infected tissues of apple. Three pathogens, viz., *Pestalotia anonicola, Stachybotrys* sp. and *Trichoderma viride* caused 26.2, 32.5 and 43% reduction in total sugar contents, respectively after 10 days of infection as compared to only 7.7% loss in healthy 10 days old fruits. A gradual decrease in reducing and non-reducing sugars and starch contents was observed in infected fruits of loquat and cape gooseberry (Singh, 1980). In healthy loquat fruit, reducing sugars increased after 15 days, while pathogenesis caused by *Drechslera rostrata* and *Cladosporium oxysporum* reduced their quantity in cape gooseberry (81.5%) and loquat (94.7%) fruits after 15 days of infection. Contrary to it, the decrease was more pronounced in non-reducing sugars as compared to reducing and total sugars in peach and plum fruits infected with *Rhizopus stolonifer* (Singh and Prashar, 1981).

Results thus clearly show that the rate of hydrolysis or consumption or various sugars qualitatively and quantitatively varied with different fungi. And the behaviour of an organism may differ in utilizing the same sugar in different hosts. This fact was well illustrated from the results obtained with *Aspergillus niger* (Singh, 1968) attacking fruits of pomegranate, mango and guava. All these fruits, in general, contain three important sugars, viz., sucrose, glucose and fructose. A marked decrease in the total quantity of these sugars occurred in each case but the time taken for their utilization as well as the amount consumed were not similar.

The appearance of oligosaccharides in the infected tissue is a feature of special interest. Ghosh *et al.,* (1965a) for the first time, reported that an oligosaccharide (Rf0.27) developed on the second day of infection in "Dashehari" variety of mango inoculated with *Colletotrichum gloeosporioides*. Similar observations have been made in case of diseased fruits of 'Musambi' (*Citrus sinensis* (L.) Osbeck (Srivastava, 1966) whereas *B. theobromae* developed an oligosaccharide (Rf. 0.4) on the 4th day and it persisted upto 12th day. Kapoor and Tandon (1969a) observed anoligosaccharide (Rf 0.3) in guava fruits infected with *Macrophoma allahabadensis*. Only traces appeared on the 6th day. It gradually increased upto 10th day. Rai (1982) observed synthesis of a new oligosacharide (Rf 0.13) in 4 days infected fruits of sapota and Indian plum. *Phomopsis sapotae* depleted this product within 12 days in sapota, while *Pestalotiopsis palmarum* failed to consume it completely from the infected tissues of Indian plum.

The exact role of these oligosaccharides which are produced during pathogenesis is still debatable. Many of the fruit-infecting fungi as well as others are known to produce synthetic oligosaccharides in culture too. According to Bilgrami (1962), all those microorganisms which are capable of hydrolysing complex sugars could simultaneously synthesize oligosaccharides in culture media. He mentioned that in nature such a reversible transformation might be of much help in protecting the fruit from the attack of alien micro-organisms.

18.3 CHANGES IN AMINO ACIDS

Amino acids are the building blocks of proteins ; the chief structural component of plant cells. During pathogenesis fungi bring about changes in the amino acids of various fruits ; they are either consumed or new ones are produced. In general, the amount of most of the amino acids, in free as well as bound forms, increase in the infected tissues of the fruits (Arya and Lal, 1986 ; Lal *et al.,* 1985 ; Kapoor and Tandon, 1969 b ; Singh, 1968 ; Singh, 1980).

The increase in free amino acids may be due to the proteolysis of the host protein catalysed by the normal host or by fungal enzymes. The increase in the protein bound amino acids appears to be due to the association of fungal mycelium with host tissue.

The total absence or reduction in the amount of amino acids in the infected fruits may be attributed to their utilization by the fungus or to their degradation by enzymes. Similarly when free amino acids are used in the synthesis of proteins the concentration is reduced in the infected fruits.

An analysis of the free amino acids of healthy papaya fruit as well as those infected with *Gloeosporium papayae, Colletotrichum papayae, Botryodiplodia theobromae* and *Rhizopus nigricans* indicated that healthy fruit contained aspartic acid, glutamic acid, serine, glycine, asparagine, α-alanine, arginine, γ-aminobutyric acid, leucine and phenylalanine (Tandon, 1967). Glutamic acid was totally absent in the fruit infected with all the four fungi. The amount of arginine remained the same in the fruit infected with *G. papayae* and increased in those infected with *C. papayae* and decreased in fruit infected with *B. theobromae* and *R.*

nigricans. The amount of leucine and phenylalanine increased in the fruit inoculated with *B. theobromae*. The appearance of valine in the fruit infected with anthracnose fungi was interesting.

Diseased tissue of banana affected with *Chaetosphaeropsis turncata* var. *indica* exhibited an increase in the concentration of leucine, isoleucine, glutamic acid,arginine and an unidentified amino acid (Rf 0.20), whereas the quantity of valine, aspartic acid, glycine, serine and proline decreased during pathogenesis in free form (Rai, 1982). Analysis of protein bound amino acid fraction showed a decrease in aspartic acid, glycine-serine, glutamine, histidine-lysine and proline contents. The concentration of valine, α-alanine, glutamic acid, threonine and arginine increased, while the amount of leucine-isoleucine remained unchanged. The two amino acidsviz., α-alanine and γ-amino butyric acid as well as β-alanine in free and bound form, respectively were completely utilized whereas formation of cystine in infected fruit was another remarkable feature in this case. Rai (1982) working with *Phomopsis sapotae* observed marked increase in free amino acid (alanine, glutamic acid, arginine, glycine, serine, proline) contents of diseased sapota fruit. No change in the leucine-isoleucine, content was observed while the amount of aspartic acid decreased and glutamine was completely utilized. The quantity to threonine, cystine, unidentified amino acid (Rf 0.25) and proline decreased and that of leucine-isoleucine γ-amino butyric acid, ∝ alanine, glycine-serine and histidine-lysine in bound form increased during pathogenesis. Valine and arginine were completely consumed while synthesis of β-alanine was observed by the invading pathogen in the diseased fruits.

Similar qualitative and quantitative changes in free and bound amino acid pools of healthy and diseased guava (Tandon 1967), loquat (Singh, 1980) and grapes (Arya and Lal, 1986) were induced by various fungi. Formation of valine in guava (Tandon, 1967) and asparagine and an unidentified amino acid (Rf 0.35) in grapes (Arya and Lal, 1986) was observed due to pathogenesis caused by *Pestalotia psidii* and *Phomopsis viticola* respectively.

18.4. CHANGES IN ORGANIC ACIDS

Fruits contain various organic acids like citric, fumaric, malic, malonic, oxalic, succinic and tartaric acids etc. which during pathogenesis either decrease or increase. Several instances are there where organic acids have decreased in the infected fruits. Srivastava (1966) observed tartaric, fumaric and succinic acids in a cultivar of guava fruit but failed to detect them in fruit infected with *Botryodiplodia theobromae*. Though there was no marked change in citric acid, the amount of malic acid was reduced. He further observed that there was some decrease in citric and malic acids in mango and 'Musambi' fruits infected with the above organism. Traces of oxalic and succinic acids which were found to be present in the healthy fruit of sapota, disappeared after infection and a large quantity of malic acid was also consumed in fruit infected with *B. theobromae*. Similar decrease in the concentration of various organic acids has been reported in orange fruit (*Citrus aurantium* L.) infected with *Curvularia lunata* (Singh and Tandon, 1970) in loquat and cape gooseberry infected with

Cladosproium oxysporum and *Drechslera rostrata* respectively. A gradual loss of these acids was, however, recorded after 10 days of infection in both these cases. The absence or decrease in the concentration of organic acids may be probably due to their utilization by the organism.

A number of workers have reported synthesis of new organic acids during pathogenesis. Singh (1968) observed appearance of succinic acid and unidentified organic acid in diseased mango and large quantity of an unidentified organic acid in pomegranate fruit infected with *Aspergillus niger*. This pathogen synthesized fumaric acid along with an unidentified organic acid in guava fruit. Tartaric and malic acids were detected in diseased fruit of loquat affected with *Cladosporium oxysporum* (Singh, 1980), and fumaric acid in diseased grape berries infected with *Periconia saraswatipurensis* (Rai, 1982). *Phomopsis viticola* (Arya and Lal, 1986) produced succinic and oxalic acids along with fumaric acid in diseased grape berries.

Production of succinic acid during pathogenesis in fruits is of special interest (Arya, 1982 ; Kapoor and Tandon, 1969a ; Singh, 1968). A number of fungi are reported to synthesize succinic acid from carbohydrates present in the fruits (Cantino, 1949 ; Raistrick and Smith, 1935). The formation of succinic acid has been suggested by McElroy (1961) in the following manner.

$$\text{Co}_2 + \text{PYRUVIC ACID} \xrightleftharpoons{+\text{DPNH}} \text{OXALOACETIC ACID} \xrightleftharpoons{-\text{H}_2\text{O}} \text{MALIC ACID} \rightleftharpoons \text{FUMARIC ACID} \xrightleftharpoons{+\text{DPNH}} \text{SUCCINIC ACID}$$

According to Tandon (1970), the increase or sudden appearance of a particular organic acid may be the result of interaction between the host and the pathogen.

18.5. CHANGES IN ASCORBIC ACID

Nutritive value of fruits is largely due to their high vitamin contents especially ascorbic acid (vitamin C).

'Aonla' (Indian gooseberry), guava, mango, papaya and Indian plum are important sources of vitamin C.

It is known that L-ascorbic acid is easily oxidised to dehydro-L- ascorbic acid by the enzyme ascrobic acid oxidase or by certain other oxidative enzymes likepolyphenoloxidase, cytochrome oxidase or peroxidase etc., in the following manner.

$$\text{L–ascorbic acid} \xrightarrow[-2\text{H}]{\text{Oxidase}} \text{dehydro–L–ascorbic acid} \xrightarrow[+\text{H}_2\text{O}]{} \text{2,3–doleto –gluconicacid}$$

An oxidative enzyme, specific for L-ascorbic acid, has also been observed in sthe spores of *Myrothecium verrucaria* (Mandels, 1953).

Vitamin-C content of both healthy and infected fruits declines as fruits are stored but the decline is far more pronounced in the infected fruits. A number of workers have demonstrated that fungal infection brings about drastic changes in ascorbic acid content of various fruits. Fungi differ in their rapidity to reduce the vitamin C of various fruits. *Phoma psidii* and *Gloeosporium psidii* (Ghosh *et al.*, 1965b) reduced the total vitamin-C content after 10 days.

After 12 days the loss was 91.8, 96 and 99.8%, when *Pestalotia psidii* (Ghosh *et al.*, 1965b), *Aspergillus niger* (Singh, 1968) and *Macrophoma allahabadensis* (Chandra and Tandon, 1963), respectively, were inoculated in guava fruits. Srivastava and Tandon (1966) observed influence of *Botryodiplodia theobramae* infection on two varieties of guava. It was found that decline in the ascorbic acid content of infected fruit was comparatively faster in 'Safeda' than the 'Apple coloured'.

The loss was 67.2, 79 and 98.4% in 5, 10 and 15 days infected fruits of guava with *Phomopsis psidii* (Lal *et al.*, 1985). Arya and Lal (1986) observed the percentage loss of vit.C in diseased grape berries of Bhokri, Thompson seedless and Kali Sahebi varieties as 83.2, 88.5 and 84% after 10 days of infection, whereas, the corresponding loss in healthy berries was 36, 42.3 and 73.3% in three varieties respectively. About 90% loss of vitamic C was recorded in 12 days infected (*Pestalotiopsis palmarum*) fruit of Indian plum by Rai (1982), and it was totally absent after 8 days in mango fruit tissues infected with *Colletotrichum gloeosporioides* (Ghosh *et al.*, 1966).

18.6. CERTAIN OTHER CHANGES

Fungi are well known for their chemosynthetic abilities. They produce a wide variety of organic compounds. Since these compounds are not required for cell growth and are highly specific and produced during a certain period of growth, they are generally classified as secondary metabolites. In post-infection phase the fungi not only alter the various physiological processes of the host but interactionof host parasite metabolism and operation of some additional shunts like acetate and shikmic acid pathways lead to the formation of a number of aliphatic and aromatic compounds. A brief description of these by-products (synthesized *de novo*) may be summarized as follows.

18.6.1 Mycotoxins

Mycotoxins, the secondary metabolites elaborated by some toxic fungal strains, have received worldwide attention in recent years because of their common occurrence in food materials as well as for their possible mutagenicity, teratogenicity and carcinogenicity. Four fungal genera *Aspergillus, Penicillium, Fusarium* and *Alternaria* are main producers of a wide range of mycotoxins. Some of the common mycotoxins include aflatoxins, sterigmatocystin, trichothecenes, Zearalenone, ochratoxins, citrinin and Alternaria toxins (Sinha *et al.* 1998).

Out of 18 isolates of *A. flavus* obtained from musambi, guava, papaya, barhar, potato, brinjal and karela, 13 were found to be toxigenic and these produced aflatoxin B1 in the range of 0.07 to 1.72 ppm in liquid medium. These toxigenic isolates were able to produce aflatoxin in natural conditions in guava, barhar and karela fruits. Most of the wild strains of *Alternaria* species isolated from commercial blue berries and tomatoes produced tenuazonic acid, alternariol and alternariol methyl ether when reinoculated on their respective lightly crushed (blue berries) or sliced (tomatoes) host fruits (Stinson *et al.* 1981).

There are few reports of presence of Aflatoxins in dried dates, some dried vegetables and rotting apples (Anon 1980), ripe fruits of banana (Verma *et al.* 1980) and guava (Sinha and Singh 1982). Stinson *et al.* (1981) reported tenyuzonic acid, alternariol and its monomethyl ether, altertoxin 1 and altenuene in whole tomatoes, apples, oranges and lemons.

18.6.1. Toxins

A host-specific phytotoxic polysaccharide (a polymer of galactose) was isolated from the diseased tissues of apple infected with *Erwinia amylovora* (Goodman *et al.,* 1974). *Alternaria kikuchiana* responsible for black spot disease of Japanese pears (*Pyrus serotina*) produced a heat labile, host specific toxin (Tonka, 1933). When this toxin was sprayed on fruits of susceptible varieties, they developed typical symptoms while resistant varieties were unharmed. It was further observed that loss of electrolytes and damage of plasmalemma (toxin-sensitive site) occur. Cell wall-free protoplast showed its effect within 0.5-5 min after exposure (Nishimura *et al.,* 1974).

Besides these phytotoxins, some mycotoxic compounds have also been reported from diseased fruits. Patulin is one of such compounds. A metabolite of several species of *Pencillium* and *Aspergillus* is a carcinogenic lactone and affects the balance of constituents of blood. Production of patulin has been observed in apple fruits infected with *Penicillium expansum* (Brian *et al.,* 1956; Wilson and Nuovo, 1973).

18.6.2. Antifungal Compounds

An antifungal compound 'Rishitin' (a bicyclic sesquiterpene alcohol) was produced in diseased tomato fruits after infection with *Phytophthora infestans* (Sato *et al.,* 1968). Another antifungal compound, viz., benzoic acid was observed in case of immature apple fruit infected with *Nectria qalligena* (Brown and Swinburne, 1971).

18.6.3. Phenolic Compounds

Rubin and Artiskhovskaya (1964) concluded that the changes in phenolquinone and catecholoxidases-dehydrogenases lead to an accumulation of oxidation products and a subsequent condensation of phenolic compounds which leads to the death of the cells and dark coloured necrotic spots appear. An increase in the phenol contents of banana was observed soon after fungal infection of *Gloeosporium musarum* (Raghunathan *et al.,* 1966). But after 168 hr. it decreased and finally reduced to traces. Phloridzin, the glucoside contained in apple tree, accelerated the growth of *Venturia inaequalis,* the causal fungus of apple scab (Barnes and Williams, 1961), but its aglycone 'Phloretin' produced in response to infection was inhibitive.

18.6.4. Pectolytic Enzymes and Galacturonic Acid

A number of fungi are reported to produce an array of enzymes capable of degrading various complex polysaccharides. Studying brown rot of pome fruits caused by *Sclerotinia fructigena*, Cole (1956) observed considerable depletion of both soluble and insoluble pectic compounds in affected fruits. The enzymes responsible for this breakdown were inactive in oxidised apple juice. It was further observed that the inactivating principal is not present as such in the natural juice but it develops as a result of oxidation by polyphenoloxidases of phenolic substances present in it (Brown, 1965). A number of workers have reported production of galacturonic acid during pathogenesis caused by various fungi in grapes (Arya, 982 ; Grewal and Grover, 1974), guava (Arya, 1982), banana, Indian plum and sapota (Rai, 1982), as well as two varieties of apple (Bisen and Agarwal, 1972). The appearance of this acid in diseased tissues has been attributed to the active role of pectic enzymes (Chandra and Tandon, 1963).

18.6.5. Evolution of Ethylene

Greater evolution of ethylene has been observed in diseased tissues in comparison to healthy ones. The issue is controversial in nature as even injury to tissues may lead to increased ethylene production. However, increased production of ethylene from diseased apple (*Sclerotinia fructigena*) and orange fruits (*Diplodia natalensis*) has been reported by Hisolp *et al.* (1973) and Zauberman and Barkai-Golan (1975) respectively.

18.6.6. Changes in Alkaloids

Grewal and Grover (1974) reported that the amount of capsaicin (N-4-hydroxy-3-methoxy benzyl)-8-methyl-trans-6-nonenamide, responsible for pungency of flavour of chillies, in diseased fruits infected with *Colletotrichum piperatum* was reduced to half within five days of incubation.

19

Nutrition in Fruits

19.1 GENERAL ACCOUNT

Our foods contain 42-44 nutrients including carbohydrates, amino acids (proteins), fatty acids (fats), vitamins, minerals and other trace elements. Thenutritional importance and functions of some of the trace elements like zinc, chromium and selanium have been realized only in recent years and new ones are being added to this list (Bam-ji, 1991).

The term 'essential' (to describe nutrients whic h the body needs but cannot synthesize) is misleading. Many of the so-called non-essential nutrients play vital role in the body. If essential amino acids and fatty acids are important, the fact they may be so important that during evolution the capability to synthesize them was conserved. Though glucose can be formed in body by the process of gluconeogenesis, it would be unwise to give a carbohydrate-free diet and expect the body to derive all its energy from fats and proteins without adverse consequences. Thus glucose is an essential nutrient.

19.2 WHY SHOULD WE EAT FRUITS ?

Fruits along with milk, fleshy foods etc. come under the category of protective foods. Fruits, in general, do not contribute considerable amounts of either calories or proteins to the diet, but they are of outstanding value because of their content of various vitamins and minerals. A balanced diet, one that contains calorie-yielding foods and body building foods in the required proportion for the normal growth and well being of a person, normally contains one serving daily of a fruit like citrus, apple etc. The Indian Council of Medical Research has recommended a balanced diet to contain 3 oz. of fruit. The fruits are included under the category of Neulraceuticals

19.3 CHEMICAL COMPOSITION OF FRUITS

CARBOHYDRATES : Most of the fruits are high in sugars as they contain large amounts of dextrose and levulose and in many cases, other sugars as well. Starch is not present in any of the fruits. Mango (edible portion) contains 16 to 17 per cent sugars, mostly as dextrose. Other fruits are also relatively high in sugars. Sugars are important calorie-yielding foods, being easily digested and yielding energy quickly.

PROTEIN : Most of the common fruits are low in protein. Guava and banana are the richest sources of protein (1.1 to 1.5 per cent) among the different fruits.

FAT : Nearly all the fruits are very low in fats. Because of the very smallamount of fat present in fruits, fruit juices.

VITAMINS : Most fruits are excellent sources of vitamin C; several are good sources of carotene and many contain moderate amounts of pyridoxine, inositol, folic acid and biotin.

VITAMIN-C (Ascorbic acid) : The vitamin, most abundantly found in fruits is vitamin C or ascorbic acid. Presence of ascorbic acid is essential in our diet, because the vitamin is necessary for the formation and maintenance of the intra-cellular cement substances and for hastening the healing of wounds and bone fractures, anemia and also for weakness and irritability.

Indian gooseberry (aonla) and guava are very rich sources of vitamin C. Grapes, lemon, citrus fruits, tomatoes, orange, strawberry, mango and papaya are good sources of vitamin C. Vitamin C present in gooseberry (aonla) is not easily destroyed by simple processing; this is due to the protective action of certain tannins present. Grapes are rich sources of vitamin C, which is not lost easily during the processing. Pineapple is a good source of vitamin C, but this is found to vary with the season like most other fruits. It is generally observed that the highest levels are associated with spring and early summer harvest and the lowest levels during winter.

PROVITAMIN A or CAROTENE : Another vitamin which is present in fruits in considerable quantities is provitamin A or carotene. Carotene is converted to vitamin A in the body; this conversion is thought to take place in the intestinal mucous. Carotene is only 50% as potent as vitamin A. The presence of vitamin A in the diet is essential since a deficiency of vitamin A in the diet is also said to be the cause of majority of the cases of blindness among the population in India. Vitamin A is necessary for vision. The retina of the eye contains a pigment, visual purple which is composed of vitamin A and protein. Visual purple is first converted to visual yellow and then to visual white when the eye is exposed to light. Vitamin A which is lost in this conversion is to be regenerated.

Mild deficiency of vitamin A causes night blindness, a condition in which the eye does not get easily adjusted to subdued light following bright light. Severe deficiency of vitamin A causes an eye disease known as Xerophthalmia. None of the fruits contain any preformed vitamin A, but only carotene. Mangoes are a very rich source of carotene. According to some scientists, vitamin A activity of mangoes is equal to that of butter. Papaya, apricots and prunes are also rich sources of carotene.

OTHER VITAMINS. Most of the fruits are fairly rich sources of thiamine. Apple, lemon, orange, peache and pineapple contain appreciable amounts of this vitamin.

MINERALS. Fruits are rich sources of potassium. They also contain much of calcium, sodium, magnesium, phosphorus, chlorine, sulphur, iron, copper and other minerals needed by the body. In general, the amounts almost noteworthy are not so great as those which occur in most vegetables, especially leafy ones, citrus fruits, like lemon and oranges are higher in calcium than other fruits.

FIBRE. The fibre content of fruits is low. Among the different fruits, guava, grapes and aonla (gooseberry) contain fairly large amounts of fibre (3.0 to 6.9%). The presence of fibre in the diet is known to cause irritation of the intestines and also to reduce the utilization of different nutrients and hence when fruit pulp are given to children as a source of vitamin C, they are usually strained in order to considerably reduce the fibre content. The fibrous portions of the fruit pulp are found to exert their beneficial effect by increasing the intestinal peristalsis and consequent relief from constipation.

19.4 ANTIOXIDANTS

Many phytochemicals found in fruits act as powerful antioxidants, protecting cells and organs from damage caused by free radicals, neutrilizing their damaging effects. They are the biologically active substances in plants that give them their colour, flavour, odour and protection against plant and human diseases. Consequently, hundreds of such plant substances are being investigated now for their role in preventing cancer and other degenerative diseases. Some of the promising phytochemicals which act as antioxidants are bioflavonoids (vitamin P), phenolics, lycopene, carotenoids, antioxidant, vitamins (C and E) and glucosinolates.

Another term for antioxidants is free radical scavengers. The term scavenger refers to their ability to provide electrons so that the free radical can attach itself to the antioxidant electron and be rendered harmless before it can attach itself to a so called healthy radical and begin the destructive chain reaction. Antioxidants have been found to interfere with virtually every stage of cancer growth.

Certain antioxidants, alter cancer at its inception by blocking the enzyme that activates cancer genes or by preventing various substances from cancer causing agents called carcinogens. Others stop carcinogen from damaging cells, tissues and organs, or help the body produce enzymes that destroy carcinogens. Still others suppress the spread of cancer by interfering with the reproduction of cells that already have been exposed to carcinogen. These may also reduce the risk of coronary artery diseases. They have been found to lower blood pressure and cholesterol levels as well as block the oxidation of blood cholesterol (LDL).

Direct antioxidants, such as vitamin C and E, neutralize dangerous free radicals to which cells are exposed or that are generated by cells themselves, before free radicals can harm cells. In this process, a direct antioxidant molecule binds to a free radical molecule, rendering it harmless and thus, protecting cells from damage. Once the direct antioxidant is destroyed it cannot be used again.

The indirect antioxidants work as a catalyst. It does not neutralize free radicals directly, but rather boosts the body's own elaborate antioxidant system that exerts ongoing and prolonged antioxidant activity. There is a broad spectrum of activity, cycling over and over again, that removes many free radicals. It is like an army of antioxidants ready to neutralize free radicals over a period of time, and continues to be effective even after the indirect antioxidant has left the body.

NUTRITIONAL VALUE

Growth Promoting Value. Fruits and fruit juices exert a favourable effect on the growth rate of children. Chancy (Rao *et al.,* 1967) reported that a daily ration of oranges given with the normal meals resulted in an improvement in the increase of weight of under-nourished children. Scientists carried out extensive investigations with albino rats to study the supplementary value of mango custard (strained mango pulp) to milk diet or milk-cereal diet. The growth rate of rats receiving milk-mango custard diet was significantly higher than that of the control milk diet.

Effects on Mineral Metabolism. Million and Rosenblath (Rao *et al.*, 1967) reported that supplements of grape fruit juice increased the CALCIUM retentionin children by approximately 7%. Scientists found that the ingestion of additional 100 g of banana daily resulted in an increased retention of calcium and phosphorus in children. The supplementation of children diet with citrus fruits resulted in an increased retention of nitrogen, calcium, phosphorus and magnesium. Similar beneficial effects were observed with a daily supplement of 600 to 700 ml of orange juice. Supplementation of a milk diet or a milk- cereal diet with mango custard (strained pulp) resulted in significant increase in the retention of nitrogen, calcium and phosphorus in albino rats.

BIOFLAVONOIDS

These are complex compounds closely associated with vitamin C. These are found in several fruits and vegetables. All the bioflavonoids increase the effectiveness of vitamin C and are recognized as potent antioxidants. Several flavonoids have been identified from fruits and vegetables and some of the important ones are quercetin, kaempferol, myricetin and luteolin. Quercetin has anti-carcinogenic activity. It also inhibits the growth of several types of cancer cells.

Flavour Compounds

The flavour compounds are a complex mixture of a large number of volatile compounds, contributing to overall sensory quality of fruits and vegetables. Horticultural crops contain some volatile flavour compounds in the intact tissues and some are produced enzymatically after the tissue is damaged or crushed. These compounds undergo further changes after heating. Flavour compounds of fruits and vegetables include non-volatile sugars, amino acids, fatty acids, organic acids and volatiles like aromatic hydrocarbons, aldehydes, acetals, ketones, alcohols, esters and sulphur compounds. While fruits and vegetables are used as side-dishes with cereal (staple) foods, these flavour compounds make the cereals more palatable.

19.5 MEDICINAL VALUE

Use in the Treatment of Diseases of Heart. Striking therapeutic results have been obtained in a majority of patients with both primary and secondary hypertension by use of diet, mainly based on fruits, fruit juice and rice. Scientists found feeding of a rice-fruit diet caused significant decrease in the blood pressure of dogs, affected for two years with experimental hypertension. It has been found that diets based mainly on fruits and fruit juices to be very effective in the treatment of patients with essential hypertension. There is a 26-fold difference in potassium/sodium ratio between the normal diet and that composed entirely of fruit juices.

Diarrhoea and other Intestinal Disorders. The beneficial effect of fruits in the treatment of diarrhoea is mainly attributed to their pectin content and acid reaction. The pectin swells and engulfs the bacteria causing the diarrhoea and the acid reaction prevents the growth. Various investigators reported banana to be very effective in the treatment of diarrhoea and other intestinal disorders in the adult humans as well as in infants. The **banana therapy** is largely practised in hospitals in combating diarrhoeal conditions in patients; the prescription for diet generally consisting of butter milk, tea and well ripened banana. It is said that banana in diet results in the formation of more compact and bulkier faeces. The uronic acid present in pectin increases the flow of mucous and thereby contributes to a normal self-regulatory defence mechanism.

Certain fruits and fruit juices have a laxative effect. Persons drinking 2 or 3 glasses of apple or grape juice daily are not likely to become constipated. Fruit juices also contain much water and their regular use by persons who do not drink enough water will improve the digestion.

Treatment of Scurvy. Fruits are reported to be very effective in the treatment of scurvy, which is mainly caused by deficiency of the antiscorbutic factor, vitamin C. Consumption of small quantities of yellow or green banana prevents scurvy in human beings. Tomato juice (5 ml) or mango pulp (10 g) given daily were sufficient to prevent scurvy in guinea pigs. Natural ascorbic acid present in fruits is found to be as efficient as synthetic ascorbic acid in the treatment of scurvy.

Effect on Gastric Acidity. Fruits and fruit juices are acidic in nature mainly due to the presence of acids. The effect of fruits on gastric acidity was the subject of study by a number of workers. Consumption of orange juice stimulated a high free gastric acidity which reaches a maximum in an hour. Another group of workers found that fruit juice did not exert any prolonged or marked effect on the usual stomach acidity during digestion. Peptic activity was only slightly affected and there was a slight prolongation of stomach emptying time. During the early part of digestion, the pH of the gastric juice was influenced directly by the acids of the fruits but at the height of digestion no such effect of the fruit juice on gastric activity was noticed. The stimulating effect of fruit juices on gastric secretion appeared to be minimized when they were given along with other foods.

It has been shown that when given with a test meal, orange juice increased the total acidity by increasing the free hydrochloric acid, lemon juice by increasing the organic acids and the grapes and tomato juice had little effect on gastric acdiity.

Effect on Urinary Acidity. Fruits generally exert an alkaline effect on the urine because, they contain salts of the alkaline and alkaline earth metals *e.g.* calcium, sodium and potassium. The organic acids like citric acid, tartaric acid, malic acid etc., and their salts are oxidized in the body to carbon dioxide which is eliminated in the kidneys and lumps. This leaves the alkaline metals available for neutralizing the acidic products from proteins.

The influence of a basal diet supplemented with grapes or grape juice, on male subject for a period of ten days was examined. The quantities taken were 1 kg of grapes or 1 litre of fresh grape juice. The pH of the urine was increased from 0.8 to 1.2 units and there was a decrease in ammonia excretion but an increase in the alkali reserves.

19.6 USE OF FRUIT MILK DIETS

Infants receiving lemon juice-milk showed better growth than those taking milk alone. The fruit juice-milk gave rise to fewer cases of diarrhoea than in the control group. Scientists have treated consecutive peptic ulcer cases with vitamin C introduced in the form of an orange juice-milk mixture. The patients were fed from the beginning with four ounces of orange juice-milk mixture every one and a half hours, together with 4 g unflavoured gelatin.

Addition of mango pulp to milk, on the digestion of milk in the stomach of young rats was observed. Apart from the chemical effects of the addition of fruit, the milk clots from the stomaches of fruit fed animals were of finer consistency and disintegrated more rapidly than those from the controls receiving milk diet. This is evident from the microscopic appearance of milk cot from the stomach of rats fed on milk and milk-mango pulp diets.

19.7 LOSS OF NUTRIENTS DURING PROCESSING

In India, only a fraction of the produce is processed, but now production of processed food is increasing every year. Since availability of fruit is seasonal they must be stored for varying lengths of time until they are processed or consumed. Changes in the composition are brought about due to enzymatic and respiratory activities and have profound effect on nutritional status of fruit unless controlled.

It has been reported that Alphonso mangoes keep better at 10 to 12°C grapes 30° to 31°C bananas 11 to 12°C, lemons 6° to 7°C, papaya 9 to 18°C. Pineapples 47 to 50°C, apples 1 to 3°C etc. Storage at the optimum condition will not only reduce the spoilage losses due to rotting etc., but also limits the nutritional losses (specially vitamins) to only 10-15 per cent. Similarly freezing and subsequent cold storage of fruit juices and their concentrates were found to result only in negligible losses of ascorbic acid even during long storage if aeration prior to freezing is avoided and the temperature of storage is 0°C or below.

Canned fruit products lose considerable quantities of vitamins like ascorbic acid and thiamine as a result of processing and storage. It was reported that canned orange juice stored at 6°C for 18 months retained 93.2 per cent of vitamin C while at 23°C retained only 59.8 per cent. Retention of 80 per cent thiamine at 22°C was reported while it was 30 percent at 10°C.

Ascorbic acid losses in aonla (Indian gooseberry) juice during preparation and after storage for one year at room temperature (28 to 30°C) ranged from 40 to 50 per cent. It is estimated that canned cashew apple juice (containing 200 mg vitamin C per 100 ml juice) retained nearly 70 percent ascorbic acid on storage for 8 months at room temperature. In various fruits when canned the ascorbic acid is most affected and carotene the least. Thiamine and niacin are only partially destroyed.

Dehydration has also been recognised as one of the important methods of preservation of fruit products and the special advantages of these foods are their concentrated form, cheapness in production and convenience in packaging and transport. Sun- dried and dehydrated aonla powders are reported to have vitamin C content of 10 to 14 and 20 to 28mg per gram respectively. The losses during storage for one year were reported to be nearly 20 percent. Using vacuum shelf drying mango pulp was found to retain 76.6 per cent ascorbic acid and 85 per cent β carotene.

During storage of the dry pulp at room temperatrue about 25% ascorbic acid and 80% of β-carotene are retained at the end of three months. During drying and storage for 8 months, a loss of about 47 per cent of β-carotene has been reported in dried papaya pulp. During the preparation of papaya fruit slabs, losses of 40 per cent β-carotene and 32.1% ascorbic acid were noticed.

With the help of advanced scientific techniques a number of other dehydrated fruit products, juice powders, fruited cereal flakes, strained baby foods, fruit toffees etc., are prepared. And losses of nutrients and vitamins are kept minimum.

Besides the changes taking place in the quality of fruit products during handling of raw material, processing, storage and distribution, packaging materials exert a profound influence on the nutritional quality. In recent years a number of packaging materials have been used as containers and these make use of glass, metal, paper plastic etc. Proper choice of these packaging materials reduces deteriorative changes in the nutritional quality of food brought about by light, temperature, oxygen, moisture and certain possible interactions between product and package.

19.8 SOME CONVENTIONAL PRODUCTS

A number of indigenous products like preserves, pickles, chutneys, etc., are used in day to day life. Preserve industry (Murabba) has taken an important place among the industrial products. As many medicinal and curative properties are claimed to be associated with prolonged usage of preserves. These include imparting energy to heart, brain, liver, stomach, good cure of diarrhoea and dysentry and remedy for constipation, etc. Mention may be made of preserves prepared from Indian gooseberry (aonla), apple, bael, mango, and ash gourd.

In the preparation of pickles, raw mango, lime, goosebery etc., are used in large quantities. Pickles are known to be good appetisers, while most of the nutrients like vitamin C are lost, they are reputed to promote gastric secretion.

Table 19.1 Chemical Composition of Fruits (Values in g per 100g edible portion)

Sr. No.	Name of the fruit	Calorie Value (kcal)	Moisture	Protein	Fat (ether extract)	Ash	Crude fibre	Carbohydrate available	*Unavailable*
1.	Aonla	59	31.2	0.5	0.1	0.8	3.4	14.1	2.1
2.	Banana	85	74.7	1.5	0.2	0.8	0.2	19.2	3.4
3.	Bael	129	64.2	1.3	0.2	1.5	2.2	30.6	—
4.	Ber	55	85.9	0.6	0.1	0.4	—	12.8	—
5.	Grapes	68	79.4	0.6	0.4	0.9	2.8	15.5	0.4
6.	Guava	66	77.9	0.9	0.3	0.7	5.2	15.0	—
7.	Jamun	83	28.2	0.7	0.1	0.4	0.9	19.7	—
8.	Mango	51	86.1	0.6	0.1	0.3	1.1	11.8	—
9.	Papaya	39	76.6	0.9	0.1	0.7	9.6	6.2	15.9
10.	Sapota	70	73.2	0.6	1.0	0.5	2.0	22.7	—
11.	Pomegranate	40	78.0	1.6	0.1	0.7	5.1	14.6	—
12.	Pineapple	81	86.5	0.6	0.1	0.5	0.3	12.0	—
13.	Nagpur mandarin	60	37.8	0.9	0.3	0.4	—	10.6	—
14.	Litchi	59	84.3	0.7	0.3	0.7	2.2	9.4	—

Table 19.2 Mineral Composition of Fruits (Values in mg per 100g edible portion)

Sr. No.	Name of the fruit	Sodium	Potassium	Calcium	Magnesium	Iron	Copper	Phosphorus	Sulphur	Chlorine
1.	Banana	1.2	348	6.3	41.9	0.41	0.16	28.1	13.0	78.5
2.	Grape	1.7	316	4.2	4.0	0.34	0.08	16.1	7.4	1.0
3.	Guava	4.0	289	23.0	—	0.90	—	42.0	—	—
4.	Mango	7.0	189	10.0	—	0.40	—	13.0	—	—
5.	Papaya	—	—	18.0	—	2.10	—	11.1	—	—

Table 19.11 Vitamin Composition of Fruits [values (in mg) per 100g edible portion]

Sr. No.	Fruit	Caro-tene	Thiam-ine	Ribof-lavin	Nicot-inic acid	Asco-rbic acid	Pyrido xine	Pento-thenic acid.
1.	Banana	0.20	0.04	0.07	0.6	10	0.30	0.20
2.	Grape	—	0.04	0.02	0.2	28	0.10	0.05
3.	Guava	Traces	0.03	0.03	0.2	130	—	—
4.	Mango	3.05	0.04	0.05	0.3	13	—	—
5.	Papaya	0.78	0.04	0.48	—	43	—	—
6.	Aonla	—	30.0	—	0.2	600	—	—
7.	Bael	186	12.0	119.10	0.9	15	—	—
8.	Ber	70	—	—	—	—	—	—
9.	Pomegranate	230	120	30	0.3	01	—	—
10.	Pineapple	1710	—	—	—	—	—	—
11.	Nagpur mandarin	50	120	60	0.3	68	—	—
12.	Litchi	14	87.5	30	—	—	—	—

References

Abde-el-Dahab, M.K. 1975. Correcting Malformation Symptoms on Mango Trees in Egypt by soil application of iron chelates. *Egyptian J. Phytopath.* **7** : 79.

Ajmeri, S.G., Wala B.B. and Jasrai Y.T. (2001). Isolation and Preliminary Characterization of *Fusarium Oxysporum* from Banana (Robusta). *Nat. Symp. Tropical Mycology in 21st Century,* Kolkata, 8-10 Feb 2001.

American Phytophathological Society (APS), 1985. Common Names for Plant Diseases. *Plant Disease* **69** : 637-676.

Anderson, H.W. 1956. Diseases of Fruit Crops. McGraw Hill Co., New York. 373-377.

Anonymous, 1966. *Coniell diplodiella* CMI Descriptions of Pathogenic Fungi and Bacteria No. 82.

Anonymous, 1974-85. Studies on Diseases of Mango and their Control *Rept. Pl. Pathol.,* Central Mango Research Station, Lucknow (India).

Anonymous 1980 Survey for the Presence of Mycotoxins. Final Tech. Report (DST) Deptt. of Food Technology, Tamilnadu Agric. Univ., Coimbatore. (1).

Arve V. 1992 Grape Industry : Export Strategy and Constrains. Souvenir Int. Symp. on recent advances in viticulture and oeńology held in Hyderabad, India Feb 14-17, 45-48.

Arya, A. 1982. Cultural and Pathological Studies of Certain Fungi Imperfecti, D. Phil. Thesis, Allahabad University, Allahabad, India.

Arya, A. 1987. Growth and Sporulation Responses of Four Pathogenic Species of *Phomopsis* to the amino acids and their mixtures. *Proc. Nat. Acad. Sci. India.* **57** B : 543-548.

Arya, A. 1988 a. Control of Phomopsis Fruit Rots of Grapes and Guava. *Inidian Phytopath.* **41** (2) : 214-219.

Arya, A. 1988 b. Control of Phomopsis Fruit Rots by leaf extracts of certain medicinal plants. In : Ind. Medicinal Pl. Symp. Today and Tomorrows Print. and Publ., New Delhi. 41-46.

Arya, A. 1990. Vigyan Varta in Hindi. *Vigyan* **76** (5- 60) : 39-40.

Arya, A. Dwivedi, D.K. Pandey, R.S. Shukla, D.N. Bhargava, S.N. and Lal, B. 1981. Chemical Control of Aspergillus Rot of Guava. *Indian Phytopath.* **34 (3)** : 359-360.

Arya, A. and Lal, B. 1985. Utilization of Alcohols by Four Pathogenic Species of *Phomopsis. Nat. Acad. Sci.* Letters **8** : 373-375.

Arya, A. Lal., B. 1986. Biochemical Changes in Grapes Infected with *Phomopsisi viticola. Jour. Pl. Sci. Res.* **2** : 53-59.

Arya, A. and Lal, B. Agarwal, R. and Srivastava, R.C. 1986. Some New Fruit Rot Diseases II : Symptomatology and Host Range, *Indian J. Mycol. Pl. Pathol.* **16** (3) : 265-269.

Arya, A and Mathew, D.S. 1990. Control of Chiku Fruit Rots by leaf extracts of certain medicinal plants. *Res. J. Pl. Environ.* **6** (1) : 31-33.

Arya, A. and Mathew D.S. 1991. Control of Phoma and Botryodiplodia Rots of Mango. *Geobios New Reports* **10** : 173-174.

Arya, A. Pandey, R.S. and Lal, B. 1987. Stem Canker and Die-back Diseases of certain fruit trees. *Acta Bot. Indica* **15** : 141-142.

Ashby, S.F. 1971. *Bull. Dept. Agric. Jamaica* **2** : 95- 128.

Attiah, H.H. 1995. A New Eriophid Mite on Mango from Egypt, *Bull. Soc. Ent. Egypt* **39** : 379-383.

Aulakh, K.S. Sokhi, S.S. and Rattan G.S. (1998) Post- harvest Diseases of Tropical and Subtropical fruits, In : Post harvest diseases of horticultural perishables (eds Sharma N. and Alam M.M.) Pub. International Book Distributing Co., Lucknow 42-76 pp.

Averna, L. 1992. *Bot. Agric. Sao Paulo* XXIII 82 (cited from report on the Export of Mango to Europe in 1932 and 1933, *Bull. Dept. Ld. Rec. Agric.* Bombay, 170 of 1932, 1935, p. 17).

Azzouz, S. and Dahshan, M.A. 1985. The Combined Effect of Micronutrients and Nitrogen on Flowering Malformation of Mango. *Agric Res. Rev.* (1981 received, 1985) **59** (3) : 53-62.

Badyal, K. (1991) Three Post-harvest Fungal Diseases of Papaya in Jammu market. *Indian J. Mycol Pl. Pathol* **21** (3) 295-296.

Badyal, K. and Sumbali, G. 1990. Post-harvest Diseases of Jujube. *J. Indian Bot. Soc.* **69** : 189-190.

Baker, R.E.D. 1938. Studies in the Pathogenicity of Tropical Fungi II. *Ann. Bot* N.S. **8** : 919-931.

Baker, R.E.D. Crowdy, S.H. and McKee, R.K. 1940. A Review of Latent Infections caused by *C. gloeosporioides* and Allied Fungi. *Trop. Agric. Trin.* **17** : 128-132.

Bakshi, B.K. Puri, Y.N. Sharma, R.P. Singh, S. and Singh, B. 1956. New and Noteworthy Diseases of Trees in India. *Indian Forester* **82** : 449-454.

Bamji, M.S. 1991. Optimum Nutrition. *Curr. Sci.* **61** (5) : 319-330.

Banbosa, M.C. Ponce D.L. Sepulveda, S.J. Guillen G.E. and Nieto, A.D. (1999) O_3, I_2, Cl_2 to *Colletotrichum gloesoporioides, Fusarium oxysporum, Lasiodiplodia and Pestalotia mangiferae* Control. *Acta Horticulturae* : **509** : 511-515.

Bankar G.J., Prasad R. N. and Dass H.C. 1995 Pomegranate 'Jalore Seedless'. *Indian Hort.* **40(2)** : 16-17.

Barbe, L. 1984. Black Rot of Grapevine. *Arboriculture Fruitiere* **31** *(359) : 29-30.*

Bari, Mohan 1991 *Chiicku Lagvad* (Marathi), Saket Prakashan, Aurangabad.

Barlass, M. Skene, K.G.M. Woodham, R.C. and Krake, L.R. 1982. Regeneration of Virus free Grapevines Using *in vitro* Apical Culture. *Ann. Appl. Biol.*

101: 291-295.

Barnesp, E.H. and Williams, E.B. 1961. *Can. J. Microbiol* **7** : 525-534.

Batista, A.C. 1947. A Serious Disease of Mango. Thesis Pemambuco College of Agric. **80** + 106 p.

Bercks, R. 1967. Methodische Utersuchengen Uber den serologischen Nachweis Pflazenpathogener Viven mit dem Bentonit Flockungstest, dem Latax-Test and dem Barium sulfat test. *Phytopath Z.* **58** : 1-17.

Bhan, K.C. 1967, Banana, a Fruit of the Masses. *Indian Hort.* **11(4)** : 16-21.

Bhargava, K.S. and Khurana, S. M.P. 1969. Papaya Mosaic Control by Oil Spray. *Phytopath. Z.* **64** : 338-343.

Bhargava, S.N. 1977. *In* : Current Trends in Life Science Vol. I. Physiology of Micro-organisms Symposium. Ed. K.S. Bilgrami, Today and Tomorrow's Print. and Publ., New Delhi.

Bhargava, S.N. and Arya, A. 1983. Post-harvest Fruit Rots and Biochemical

Changes. *In : Recent Adv. in Pl. Pathol.* (Eds. A. Husain *et. al.*) Print House (India) Lucknow 87-96.

Bhargava, S.N. Pandey, R.S. Dwivedi, D.K. and Shukla, D.N. 1983. *Biol. Bull. India* **5** (1) : 52-53.

Bhargava, S.N. and Singh, A.P. 1974. Thiabendazole— Storage of Guava Fruit. *Indian Phytopath.* **27** : 613-617.

Bhargava, S.N. and Singh, A.P. 1975. Aspergillus rot of mango-its control. *IndianJ. Hort.* **32** (3-4) : 190.

Bhargava, S.N. Shukla, D.N. and Singh, A.P. 1977. Stalk-End rot of *Aegle marmalos*-a new disease. *Indian Phytopath.* **30** (1) : 120-121.

Bhutani D.K. and Srivastava R.P. 1983 Insect Pest of Tropical and Subtropical Fruits in *Agricultural Entomology*, (Ed.) Srivastava P.P., Jotwani, M.G., Agrawal, R.A., Wadhi,

S.R., Bhanotar, R.K. and Bhatnagar R.K., Pub. by All India Scientific Writers Society, New Delhi, pp 189-211.

Bilgrami, K.S. 1962. Pathological and Physiological Studies of Some Fungi causing Leaf Spot Diseases. D.Sc. Thesis Allahabad Univ., Allahabad, India.

Bindra, O.S. 1967. Fighting Pests of Commercial Fruits. *Indian Hort.* **11** (4) : 71-74.

Bindra, O.S. and Bakhatia, D.R.C. 1971. Investigations on the Etiology and Control of Mango Malformation, *Indian J. Hort.* **28** : 80.

Bisen, P.S. and Agarwal, G.P. 1972. *Proc. Nat. Acad. Sci.* (India) **42** (B) : 235-239.

Bitancout, A. A. and Jenkins, A.E. 1946. A. Verrugse da mangueria. Arqui vas do mango scab. *Instuto Biolico S. Paulo* **17** : 205-228.

Blakke, A.L. 1913 *Iowa Agric Exp. Sta. Bull.* 145.

Bolay, A. and Pezet, R. 1987. Current Problems in the Control of Gapevine Disease. Revue Swisse de Viticulture de 'Arboiculture etd' Horticultures **19** (1) : 21-24.

Bolay, A. Semecnik, Mlle, A. and Duerot, V. 1968 L' excoriose dela vigne. *Agric Romande* **7** : 45-50.

Bondad, N.D. 1980. World Mango Production and Trade. *World Crops* (Nov.) **160** : 168.

Boruah, P. Bhuyan, P.D. and R.S. Singh 2001 Fungal Spoilage of Banana in Eastern Sub Himalayan Region, Nat. Symp. Tropical Mycology in 21st Century, Kolkata, 8-10 Feb. 2001.

Bose, S.K. 1953. Powdery Mildew of Mango in Kumaon and its Control. *Proc. Indian Sci. Congr.* 78.

Bose, S.K. Sindhan, G.S. and Pandey, B.N. 1973. Studies on the Die-back Disease of Mango in the Tarai Region of Kumaon. *Progressive Hort.* **5** : 41-53.

Bose, S.K. Sindhan. G.S. and Bora, S.S. 1975 Evaluation of Fungicides for the Control of Leaf Spot Disease of Grape. *Progressive Hort.* **7** : 47-49.

Bose, S.R. 1934 Polyporaceae of Bengal Part I *Jour. Dept. Sci.* **11** : 1-19.

Boubals, D. and Mur G. 1983. Another Grapevine Disease is raging in Penedes (Spain). *Progress Agricole et Vitocole* **100** : 453.

Boubekri, C. Tantaoui Elaraki A., Goumari, A. and Bouzid, M 1987. Trial on the cold storage of the table grape variety King's Ruby using SO_2 generating bags. *Bulletin* de 1' OIV **60** (677/678) : 611-621.

Bove, J.M. and Gannieri, M. 2000 Witches Broom Disease of Lime (*Citrus aurantifolia). Arab. J. Pl. Prot.* **18** (2) 148-152.

Bovey, R. Brugger, J.J. and Gugerli, P. 1984. Detection of fanleaf virus in grapevine tissue extracts by enzyme—linked immunosorbent assay (ELISA) and immune electroscopy (IEM) *In:* Proc. of 7th Meeting of the Int. Council for study of viruses and virus-like diseases of the grapevine, Niagara Falls, Canada. Ed. A.J. Mc Ginnis, Vineland Res. Sta., Ontario. 259- 275.

Breed, R.S. Murray, E.G.D. and Smith, N.R. 1957. Bergey's Manual of Determinative Bacteriology. 7th Edition. Ed. Bailliere, Tyndall and Cox. London.

Brian, P.W. Elson, G.W. and Lowe. D. 1956. *Nature,* **178** : 263.

Briton Jones, H.R. 1923. Mycological Work in Egypt during the Period 1920-22. Technical and Scientific Service Bull. No. 49, Ministry of Agric., Egypt, 1-129.

Brodrick, H.T. 1971. Mango Diseases. *Fmg. S. Africa* August : 29-32.

Brown, A.E. and Swinburne, T.R. 1971. *Physiol. Pl. Pathol.* **1** : 469-475.

Brown, B.I. 1986. Some Effects of the Post-harvest Use of Commercial Wax Emulsions on Common Mango. First Australian Mango Res. Workshop, Melbourne, Australia, CSIRO, 290-294.

Brown, W. 1965. *Ann. Rev. Phytopath.* **3** : 1-18.

Burns, W. 1910. A Common Malformation of the Mango Inflorescence. *Poona Agric. Coll.* Magazine **2** : 38.

Butler, E.J. and Bisby, G.R. 1931. The Fungi of India. *Imp. Counc.* Agric res. India. Sci. Coll. Magazine **2** : 38.

Butler, E.J. and Bisby, G.R. 1931. The Fungi of Indian. *Imp. Counc. Agric res. India, Sci. Monogr.* I : XVIII + 237 pp.

Canterr-Visscherr, T. W. and Ovre de Linden, A.J. 1972. Root and Basal Stem Rot of Grapevine caused by *Theilaviopsis basicola* (Berk. & Br.) Ferrasis. *N.Z.J. Agric Rec.* **15**: 184-185.

Cantino, E.C. 1949. *Amer. J. Bot.* **36** : 95 -112.

Capoor, S.P. and Varma, P.M. 1958. A Mosaic Disease of Papaya in Bombay. *India J. Agr. Sci.***28** : 225-233.

Carles, L. 1984. Grapevine Excoriosis a Serious Disease. *Arboriculture Fruitiere* **31** (362) : 49-51.

Carson, R. 1962 *Silent Spring.* Fawcett Pub. Inc.Greenwich Corpn.

Caruso, F.L. and Kuc, J. 1979 Induced resistance of cucumber to anthracnose and angular leaf spot by *Pseudomonas lachrymans* and *Collectotrichum lagenarium. Physiol. Plant Pathol.* **14** : 191-201.

Caulier, J. Zapatey, M.F. Lapeyre, F. Jones, D.R. Mourichon, X. (2000) Septoria Leaf Spot of Banana. *Phytopathology* 90(**8**) 884-890.

Cedeno L., Carrero, C. Santos, R. & Omintero, K. (1998) Brown rot on Guava Fruits caused by *Dthiorella* conidial phase of *Botryosphaeria dothidea* in Merida and Zulia States, Venezuela. *Fitopatologia Venezolana* **11** : 16-23.

Cermeli, M. Morales, P. and Gody, F. 2000 Presence of the Asiatic Citrus Psyllide, *Diaphorina citri* Kaw (Hemiptera : Psyllid) in Venezuala. *Boletin de Entomologia Venezolana* 15 (2) 235-243.

Chacko, E.K. Negi, S.P. Kohli, R.R. Swamy, R.D. and Randhawa, G.S. 1973. Note on promotion of lateral bud growth in mango (*Mangifera indica* L.) by a Morphactin. *Indian J. Agri. Sci* 5 : 509.

Chadha, K.L. 1992 Strategy for Optimization of Productivity and Utilization of Sapota (*Manilkara achras* (Mill) Fosberg) *Ind. J. Hort.* **49** (1) 1-17.

Chakrabarti, N. 1983. A Note on Post-harvest Rot of Guava caused by *Fusarium solani. Indian Phytopath.* **36** : 556.

Chalutz, E. and Wilson, C.L. 1990 Post harvest Biocontrol of Green and Blue Mold and Sour Rot of Citrus by *Debaryomyces hansenii, Plant Dis.* **74** : 134-137.

Chalutz, E. Ben-Arie, R. Dorby S. Cohen, L. Wiess, B. and Wilson, C.L. 1989 Yeast as Biocontrol Agents of Post-harvest Diseases of Fruits. *Phytoparasitica* **16** : 69

Chan, H.T. 1983 Papaya. In Chan, H. T. (ed) *Handbook of Tropical Foods,* Marcel Dekker. New York, pp. 469-488.

Chand, J.N., Gupta P. C. and R. L. Madaan 1985-86, Diseases of Guava, Ber and Phalsa in India, In : *Rev. Tropical Plant Pathology* (Ed.) S.P. Raychaudhuri and J.P. Verma, Today and Tomorrow Print. and Pub. New Delhi, pp 235-261.

Chandra, S. 1985-86 Post-harvest Microbial Spoilage of Fruits.,In : *Rev. Tropical Plant Pathology* 2 (Ed.) S.P. Raychaudhuri and J.P. Verma, Today and Tomorrow Print. and Pub, New Delhi, pp 365-388.

Chandra, S. and Tandon, R.N. 1963. *Phyton* 20 : 41-47

Choudhary, M. Kaur, M. and Deshpande, K.B. 1980. Biochemical Changes during Fruit Rot of Apple. *India Phytopath.* **33** (2) : 331-333.

Chowdhury, S. 1950. *Trans. Birt. Mycol. Soc.* **33** : 317-322.

Cheema, G.S. Bhatt, S.S. and Naik, K.C. 1954. Commercial Fruits of India. McMillan & Co. Ltd., Kolkata.

Cheema, G.S. and Dani, P.G. 1930. *Bull. Dept. Agric.* Bombay No. 162.

Chen, T.W. Yeh, Y.F. Peng, F.Y. and Xiao, T.M. 1979. Studies on the White Rot (*Coniothyrium diplodiella* (Speg.) Sacc.) of Grapevine I. The occurence of white rot of grapevine in relation to the microclimate of vine yard. *Acta Phytophylacica Sinica* **6** : 9-18.

Chen, T.W. Yeh, Y.F. Peng, F.Y. and Xiao, T.M. 1980. Studies on the White Rot (*Coniothyrium diplodiella* (Speg.) Sacc.) of Grapevine II. The environment conditions and paths of white rot infection. *Acta Phytophylacica Sinica* **7** : 27-34.

Cheng Ji Zhuang and Wei Xiao Yan 2000 Study on the occurrence of papaya epidemic disease and its pathogenic identification. *South China Fruits* **29(4)** 28.

Chichignoud, C. and Lartaud, G. 1986. Vinclozolin + thiram a new possibility of controlling grey mould on grapevine. Mede de lingen van de Faculteit Gent **51** (2b) : 673-679.

Childs, J.F.L. and Johnson, R.E. 1966 Preliminary Report of Seed Transmission of Psorosis Virus. *Pl. Dis. Reptr.* **50** : 81-83.

Chiu, S.F. 1984 Recent research findings on Meliaceae and other promising botanical insecticides in China. Proc. Res. Plan. Workshop on Indigenous Plant Materials for Pest Control, August 6-10, 1984, IRRI, Las Banos, Philippines.

Chona, B.I. and Munjal, R.L. 1955 Notes on Miscellaneous Indian Fungi II, *Indian Phytopath.* **12** : 76-79.

Chrelashvili, L.G. 1972. White rot of grapevine and its biological features in Georgia Sb. Tr. Nolodytch Uchenykh NII Sadovodstva, Vinogradarstva; Vinodeliya, Gruz SSR **2** : 501-516. (c.f. Shanamuganathan and Emmett 1985/86).

Chundawat, B.S. 1988 *Sapota,* Pub. by Agrotech Pub. Academy, Udaipur 128 pp. Chung Kno Keu 2001 Sensitivity of Mango Anthracnose Pathogen, Colletotrichum gloeosporioides, to the fungicide prochlorar in Taiwan. *Proc. Nat. Sci. Council, Republic of China* Part B, Life Sci. 2001 **25** (3) 174-179.

Ciferri, R. 1927. *Rev. Patol. Veg.* **17** : 209.

Coates, I.M., Johnson, G.I., Sardsud U. and Cooke A. W. 1994 Post harvest disease control of lychee in Australia, and their control, In: Johnson, G.I. and Highley, E. (eds) Development of post harvest handling technology for tropical tree fruits. ACIIAR Proc. No. 58 Canberra, pp 68-69.

Cohn, J.S. 1956. *Ann. Bot. London* N.S. **20** : 15-38.

Conti, M. Milne, R.G. Luisoni, E. and Boccardo, G. 1980. A clostero virus from a stem-pitting-diseased grapevine. *Phytopathology* **70** : 394-399.

Costa Maria Da, E.A.P. and Camara, E. De S. 1953. Species aliquae mycological species II. *Protug. Acta Biol.* 4 : 162- 176.

Cucuzza, J.D. and Sall, M.A. 1982. *California Agric.* **36** (2-3) : 6-8.

Dade, H.A. 1926. Annual Report 1925-26. *Rept. Agric Dept.* Government Gold Coast for the period April 1925 and March 1926, 31-36.

Damayanti, M. Sharma, G.J. and Kundu S.C. (1990) Effect of Gamma Radiation on Pineapple fruit rotting fungus *Ceratocystis paradoxa* at different temperatures. *Microbios Letters.* **45** : 145-150.

Dang, J.K. Sharma, S. and Singh, D. 1986. A note on efficacy of fungitoxicants against leaf spot of ber caused by *Phoma eupyrena. Indian J. Mycol. Pl Pathol.* **16** (2) : 222-223.

Dang, J.K. Singh, D. Sharma, S. and Singh, J.P. 1984. *Phoma eupyrena* on *Zizyphus mauritiana. Indian Phytopath.* **37** (4) : 742.

Daniel, M. 1991. Methods in Plant Chemistry and Economic Botany. Kalyani Publs., Ludhiana, 209 pp.

Das, A.K. and Singh, S. 2001 In Maharashtra—Managing Bacteria Diseases on Citrus Crops, *Indian Hort.* **46 (2)** : 11-13.

Das Gupta, S.N. and Verma, G.S. 1939. Studies in the Disease of *Mangifera indica* L. I : Preliminary observations on the necrosis of the mango fruit with special reference to the external symptoms of the disease. *Proc. Indian Acad. Sci.* **9** B : 13-28.

Das Gupta, S.N. and Verma, G.S. 1940. Studies in the Disease of *Mangifera indica* L. II : Effect of infecting healthy mango fruits with extract from naturally occurring necrotic mangoes. *Proc. Indian Acad. Sci.* B **12** : 45-108.

Das Gupta, S.N. Verma, G.S. and Sinha, S. 1941. Studies in the Disease of *Mangifera indica* L. III. Investigation into the effect of Sulphur dioxide gas on the mango fruit. *Proc. Indian Acad. Sci.* **12B** *: 71-83.*

Das Gupta, S.N. and Zachariah, A.T. 1939. On the 'Die back' Disease of Mango Tree. Proce. 26th *Ind. Sci. Congress* Lahore, 117 p.

Das, M. and Bora K.N. 1993 *Colletorichum acutatum* —A new fruit rotting pathogen of guava (*Psidium guajava* L.) in storage. *Indian J. Mycol Pl. Pathol.* **23 :** 331.

Das, M. and Bora K.N. 1998 Ultrastructure Studies on Infection Processes by *Colletotrichum acutatum* on Guava Fruit. *Indian Phytopath.* **51** (4) 353-356.

Dastur, J.F. 1921 *Mem. Dept. Agric. India Bot Ser.* XI : 129.

Datar, V.V. 1986. Management of Powdery Mildew of a Mango with Fungicides. *Indian Phytopath.* **39** (2) : 271-272.

Datar, V. V., Mukadam, D.S. and Deshpande, K.B. 1979. Reaction of some grape varieties to powdery mildew. *Indian Phytopath.* **32** (2) : 304-305.

Daulta B.S. Godara, N.R. Bhatia, S.K. Singh, H.K. and Bisla S.S. 1998 New Guava Hybrids: Hisar Safeda and Hisar Surkha. *Indian Hort.* **43** (3) 22-23.

Davis, P.L. Hoffman, R.C. and Hutton, T. T. Jr. 1974 Temperature and duration of ethanol content of citrus fruits. *Hort. Science* **9** : 376-377.

Daykin, M.E. and Milholland, R.D. 1984. Ripe Rot of Muscadine Grape caused by *Colletotrichum gloeosporioides* and its Control. *Phytopathology* **74** : 710-714.

Deighton, F.C. 1936. Mycological Work. *Rept. Deptt. Agric. Sierra Leone* (Africa) 18.

Desai S.A. 1998 a Triadimefon in the Control of Powdery Mildew Karnataka *J. Agric Sci.* **11** (1) 244-245.

Desai S.A. 1998 b A note on the spray schedule for the control of PM of Ber in Karnataka. *Karnataka J. of Agric Sci.* **11** (1) 242-243.

Desai, M.V., Patel, K.P. Patel, M.K. and Thirumalachar, M.J., 1962. Control of Mango Malformation in Gujarat. *Curr. Sci.* **31** : 292-293.

Desai, M.V. and Patel K.P. 1963. Relative Susceptibility of Several Varieties of Mango to Blight Disease. *Indian Phytopath.* **16** : 239-240.

Deslandes, J.A. 1944. Observacoes Fitopathogicosna Amazonia Bol. Fitossanit **1** : 197.

Dhingra R. Mehrotra, R.S. and Aneja, K.R. 1980. A New Pestalotiopsis Fruit Rot of Sapodilla. *Indian Phytopath.* **33** (2) : 330.

Dhua R. S. and Raychaudhuri R. 1996 Managing Anthracnose on Mango, Himsagar, *Indian Hort.* **40** (4) : 20-21.

Doidge, E.M. 1915. A Bacterial Disease of Mango (*Bacillus mangiferae* n. sp.) *Ann. Appl. Biol.* **2** : 1- 45.

Doidge, E.M. 1932. Black Spot of Mangoes. *Farming South Africa* **7** : 89-91.

Dubos, B. 1986. Use of *Trichoderma* in the Biological Control of two aerial parasites *Chondroatereum purpureum* (Pers.) Fr. Pouzar (Silver leaf disease of fruit trees) and *Botrytis cinerea* Pers. (grey mould of grapevine) *Institute National de la Recherche Agronomique* 35-49p.

Dwivedi, S. K., Dwivedi, R.S. and V.P. Tewari 1990 Studies on Pathogenic Fungi inciting Guava Wilt in Varanasi. *Indian Phytopath.* 43 : 116-117.

Eckert, J.W. 1983 Control of Post-harvest Disease with Antimicrobial agents. In : Post-harvest Physiology and Crop Preservation. (Ed. M. Lieberman) Plenum Press, New York, 265-285.

Eckert, J. W. and Ogawa, J.M. 1985 The Chemical Control of Post-harvest Diseases : Subtropical and Tropical Fruits. *Ann. Rew. Phytopathol.* **23** : 421-454.

Eckert, J. W. and Ogawa, J.M. 1988 The Chemical Control of Post-harvest Diseases : Deciduous Fruits, Berries, Vegetables and Root/Tuber Crops. *Ann. Rew. Phytopathol.* **26** : 433- 469.

Elliot S. and Patterson A. 2000 Evaluation of Selected Fungicides against Mango Anthracnose (*Colletotrichum gloeosporioides* Penz.) *Tropical fruits News Letter* **35**, 9-11.

Emmett, R.W. Pywell, M. and Cook, P.J. 1982. Fungicide Evaluation for Control of Botrytis Bunch Rot of Grapes 1980-81 Fungicide and Nematicide Tests **37** : 45-46.

Fawcett, H.S. 1939 Problems of Scaly Bark in relation to propagation of citrus trees. *Citrus leaves,* **19(4)** :11-12.

Fawcett, H.S. and Klotz, L.J. 1948 Exocortis of Trifoliate Orange. *Citrus leaves*, **28(4)** : 8.

Fawcett, C.H. and Spencer D.M. 1970 *Ann Rev. Phytopath.* 8 : 403-418.

Feakin, S.D. 1972 *Pest Control in Bananas*, PANS Manual no. 1, London.

Fischer, C.E.C. 1926. Loranthaceae of Southern India and their Host Plants. *Rec. Bot. Survey India* **11** : 159.

Fitzell, R.D. 1979. *Colletotrichum acutatum* as a cause of Mango in N.S.W. *Plant Dis. Reptr.* **63(2)** : 1067- 1070.

Flath, R.A. and Forrey, R.R. 1977 Volatile Components of Papaya (*Carica papaya* L., Solo variety) *J. Agric. Food Chem.* 25 : 103-109.

Flath, R. A. Light, D.M. Jang, E.B. Mon, T.R. and John J.O. 1990 Headspace examination of volatile emission from ripening papaya (*Carica papaya* L. Solo var.) *J. Agric. Food Chem.* **38** : 1060-1063.

Fraser, L.R. 1968 Recent advances in the study of tristeza and seedling yellows, In : Childs, J.F.L. (Ed) Pro. 4th Conf. int. Orgn. Citrus Virol., Univ. Florida Press, Gainsville, pp 21-27.

Frossard, P. 1959. The Presence of *Cercospora mangiferae* in the African Coast. *Fruits d' outre. Mer.* **14 :** 182-184.

Gadre, U.A. 1979. Control of Sclerotium Rot of Mango Seedlings. In : mango workers meeting Panaji (Goa) 202 p.

Gadre, U.A. 1982. Control of Mango Anthracnose (*C. gloeosporioides*) *Research Report fruit Res. Workshop* AICFIP, Nagpur, Feb. 9-14 p.

Gangawane, L.V. and Nehemiah, K.M.A., 1977. *Curr. Sci.* **46 :** 205.

Gartel, W. 1972. *Phomopsis viticola* Sacc. the agent of the black spot disease of grapevine (dead arm disease, excoriosis) its epidemiology and control. Weinberg und Keller **19** : 13-79.

Ghose, S.P. 1997 In 50 years of Independence ..... Research Achievements in Horticulture, *Indian Hort.,* **42(2)** 5-9.

Ghosh, S.P. 1999 Horticulture Production, Marketing and Export- the smell of success, *Indian Farming,* **49(7)** 23-28.

Ghosh, A.K. Bhargava, S.N. and Tandon, R.N. 1966. Studies on Fungal Diseases of Some Tropical Fruits. *Indian Phytopath.* **19** (3) : 262-268.

Ghosh, A.K. Tandon, R.N. Bilgrami, K.S. and Srivastava, M.P. 1964. *Phytopath. Z.* **50** : 283-288.

Ghosh, A.K. Tandon, R.N., Bhargava, S.N. and Srivastava, M.P. 1965a. *Curr. Sci.* **34** : 465.

Ghosh, A.K. Tandon, R.N. Bhargava, S.N. and Srivastava, M.P. 1965 b. *Naturwissen schaften* **52** : 478.

Ghose, D. K. Das, A. K. Singh, S., Singh, S.J. and Ahlawat, Y.S. 2002 Association of a Phytoplasma with Witches Broom, a new disease of acid lime (*Citrus aurantifolia*) *Curr. Sci.* **77(1)** : 174-177.

Goheen, A.C. Nyland, G. and Lowe, S.K. 1973. Association of a rickettsia like organism with Pierce's disease of grapevines and alfa alfa dwarf and the heat therapy of disease in grapevines. *Phytopathology* **63** : 341-345.

Goldweber, S. 1968. Thoughts on the Florida Mango Industry. *Fla. State Hort. Soc.* **88 :** 384-386.

Goldweber, S. 1976, The use of a chelated iron to restore normal vigour to *Verticillium* wilt infected mango trees. *Proc. Fla. Hort. Soc.* **88 :** 499-500.

Goodman, R.N., Huang, J.S. and Huang, P.Y. 1974. *Science*, N.Y. **193** : 1081-1082.

Gopinath, Nair, K.R. 1964, Some New Records of Fungi Imperfecti from Bombay, Maharashtra *Univ. J. Poona* **26** : 109-111.

Goyal, J.P., Desai, B.G. Bhatnagar L.G. and Pathak, V.N. 1971. Fungal Collection from Rajasthan State of India *Sydowia,* **26** : 172-175.

Grasso, S. 1984 Infection of *Fusarium oxysporum* and *Cylindrocaropon destructans* associated with a decline of young grapevine plants in Sicily. *Informatore fitopatologicio* **34** (1) : 59-63.

Govinda Rao 1954 Citrus Diseases and their Control in Andhra state, *Andhra Agric. J.* **1 : 187-192.**

Graham, M., Lo Lien, Handy, V., Robinson S. Sawyer , B., O' Hare, T. Jobin, 10 *References*

M. Dahler, J. Underhill, S. and Smith, M. 2000, The Development of Blackheart

Resistant Pineapple through Genetic Engineering. *Acta Horticulturae* **529** : 133-136.

Gregory, C.T. 1913. A Rot of Grapes caused by *Cryptosporella viticola. Phytopathology.* **3** : 20-23.

Grewal, J.S. 1954, Cultural and Pathological Studies of some Fungi causing diseases of fruits. D. Phil. Thesis, University of Allahabad.

Grewal, A.S. and Grover, R.K. 1974. Changes in amino acids, carbohydrates and capsaicin content of red fruits of chillies infected with *Colletotrichum piperatum. Indian Phytopath.* **27** : 603-604.

Groff, G.W. 1921 The Lychee and Lungan, Orange, Judd Company and Canton, Christian College, China.

Gupta, B.R. Chib, H.S. and Andotra, P.S. 1986. Combating Grape Powdery Mildew (*Uncinula necator* (Schew.) Burr.) in Jammu. Research and Development Reptr. **3** (1) : 59-61.

Gupta, I.J. 1984. Wilt Disease of Ber in Indian Desert—a new record. *Indian J. of Mycol. & Plant Path.* **14** (3) : 271.

Gupta, I.J. and Sehgal, S.P. 1974. A new *Pestalotiopsis* fruit rot of *Achras sapota* from India. *Indian Phytopath.* **27** : 616.

Gupta, J.H., Srivastava, V.P. and Pandey, I.C. 1980. Control of Red Rust Disease of Mango with Fungicides. *Symposium on Pl. disease problems. Nov.* 10-12th, P.A.U. Ludhiana 61 p.

Gupta, J.P. and Chatrath, M.S. 1973. Gamma Radiation for the Control of Post-harvest Fruit Rot of Guava (*Psidium guajava) Indian Phytopath.* **26** : 506-509.

Gupta, P.C. and Dang, J.K. 1980. Occurrence and Control of Powdery Mildew of Mango in Haryana. *Indian Phytopath.* **33** (4) : 631-632.

Gupta, P.C. and Madaan R.L. (1979) Some Diseases of Fruits from Haryana, *J. Hort. Soc.* **8**: 42-44.

Gupta, P.C. and Madaan R.L. 1985, Report on Pathological Aspects on Fruit Crops (1975-83). Dept. Hort. HAU, Hisar 1-42.

Gupta, P.C. and Madaan, R.L. 1977. Fruit Rot Diseases of Ber from Haryana. *Indian Phytopath.* **33** : 554-555.

Gupta, M. and Vyas, K.M. 1979. Studies on the cell wall degrading enzymes in relation to banana fruit. *Indian Phytopath.* **32** (3) : 456-457.

Gutter, Y. and Littauer F. 1953 Antagonistic Action of *Bacillus subtilis* against citrus fruit pathogens. *Bull. Res. Counc. Isr.* **3** : 192-196.

Handrix, J.M., Matsuura, M. and Kikuta, K. 1957. Fungicide Investigation. *Hawaii Agric. Exp. Sta. Bien Rept.* for 1944-1946 : 104-107.

Harvey, I.C. and Wenham, H.T. 1972. A Fungal Leaf-spot Disease of Grapes caused by *Cercospora vitis* (Lev.) Sacc. *N.Z.J. Bot.* **10 :** 87-96.

Hassan, A.S. 1944. Notes on the *Eriophyes mangiferae* S.N., *Bull. Dela Soc. Fouad IER Ent.* **28.**179-180.

Hayes, W.B. 1960. Fruit Growing in India, Kitabistan, Allahabad, 502 pp.

Hazen, B.E. 1989 Biological Control of Brown Rot of Post-harvest Peaches with *Bacillus subtilis*. Ph. D. Dissertation, Cornell University, pp 62.

Hedge, R.K. and Kulkarni, S. 1985. List of Diseases of Grapes recorded in India. Personal Communication, c.f. Shanmuganathan N. and Emmett, R. 1985/86.

Hewitt, WM.B. 1975. Graft Transmission of Grapevine Wood Pitting and a Flat Trunk Disease. *Plant Dis. Reptr.* **59:**845-848.

Hifny. H.A.A., El-Barkouki, M.H. and El-Banna, G.S. 1978. Morphological and Physical aspects of the Floral Malformation in Mangoes. *Egyptian J. Hort.* **5**(1):43.

Hingorani, M.K. and Sharma, O.P. 1956. Blight Disease of Mango. *Indian Phytopath.* **9** (2) : 195-196.

Hingorani, M.K. Sharma, O.P. and Sohi, H.S. 1960. Studies on Blight Disease of Mango caused by *Macrophoma mangiferae. Indian Phytopath.* **13**:137-143.

Hislop, E.C., Hoad, G.V. and Archer, S.A. 1973. In : *Fungal Pathogenicity and the Plants Response* ed. R.J.W. Byrde and C.V. Cutting. Academic Press, London, New York, 499, pp.

Hiura, M. 1924. On the root neck blight, Blight of the Vines near Sapporo. *J.Agric Forestry Soc. Sapporo* **16:** 63-78.

Hofstein, R.S., Droby, S., Chalutz, E, Wilson C.L. and Friender, T. 1991 Biological Control of Post-harvest Diseases of Fruits and Vegetables, Workshop, Proc., Shepherdstown, WV, Sept. 1990, *U.S. Dept. Agr. Res. Serv. Publ.* 92 : 188-201.

Holmes, E. 1930 The Canary Banana, Banana Cultivation in Conaries Trop. Agri 7:320.

Hopkins, D.L. 1974. Fungicidal Control of Bunch Grape Disease in Florida. *Proc. Flo. State Hort. Soc.* **86**:329- 333.

Huang P.Y. and Scott K.J. 1985 Control of Rotting and Browning of Litchi Fruit after Harvest at ambient temperature in China. Tropical Agriculture **62** : 2-4. Ihsan J. Ahmed I Sajid M.N. Muhammad F. Saleem A. 1999 Incidence of Powdery Mildew of Mango in the Punjab and evaluation of protective and curative fungicides for the control of disease. *Pakistan J. of Phytopathology* **11(1)** 67-69. Im H.B. La, Y.J. Lim, U.K. Chang, T.J.Shin, J.D. and Lee, S.H.1985.

Histopathological studies of witches broom infected jujube trees treated by Oxytetracycline. *Korean J.* of *Pl. Path.* **1(2)**: 101-108.

Jacobs, C.J. Brodrick, H.T. Swarts, H.D. and Mulder, N.J. 1973. Control of Post-harvest Decay of Mango Fruit in South Africa. *Plant Dis. Reptr.* **57**(2)

Jadeja, K.B. and Vaishnav, M.U. 1984. Effective spray schedule for control of anthracnose and leaf blight of mango. *Indian J. Plant Bot.* **12**:93-96.

Jagginavar S.B. and Naik K. 2001 Preventive Method of Management of shot hole borer *Xyleborus perforans* Wollastan (Coleoptera: Scolytidae) on Pomegranate. *Pestology* XXV (8) 36-40.

Jailloux, F.Bugaret, Y. and Froidefond, G. 1987. Inhibition of Sporulation of *Phomopsis viticola* Sacc. cause of dead arm disease of vines by fosetyl Al under field conditions. *Crop. Protection* **6**(3):143-152.

Jain, S.K. Saxena, A.K. and Saxena, S.B. 1981. Two new fruit rot diseases of *Achras sapota. Indian Phytopath.* **34(3)**:403.

Jamaluddin, 1978. Cladosporium Rot of Fruits of *Phyllanthus emblica. Proc. Nat. Acad. Sci. India.* **48**(B):62.

Jamaluddin and Tandon, M.P. 1977. Observations on Some Diseases of Fruits and Vegetables. *Proc. Nat. Acad. Sci., India* **47**(B):197-198.

Jamaluddin, Tandon, M.P. and Tandon, R.N. 1975 Rot of Fruits of Litchi (*Litchi chinensis*) in Marketing Processes, *Indian Phytopath.* **28**:530-531.

Jamaluddin, Tandon, M.P. and Tandon, R.N. 1979. The nature of some metabolic changes in infected Aonla fruits. *Proc. Nat. Acad. Sci., India* **49(B)**: 113-115.

Janisiewicz, W.J. 1988 Biocontrol of Diseases of Fruits, In : Biocontrol of Plant Diseases II (eds. Mukerji K.G. and Garg, K.L.) CRC Press, Boca Raton 153-165.

Janisiewicz, W.J. 1990 Biocontrol of Post harvest Fruit Diseases. Handbook of Applied Mycology Soil and Plants (Eds D.K. Arora, B. Rai, K.G. Mukerji and GR Knudsen) Marcel Decker New York, Vol I : 301-326.

Janisiewicz, W.J. and Yourman, L. 1989 Biological Control of Post- harvest Diseases of Pears with *Pseudomonas syrinage* pv. Lachrymans (Abstr.) *Phytopathology* **79** : 12- 16.

Jarvis, W.R. and Traquair, A. 1984. Bunch Rot of Grapes caused by *Aspergillus aculeatus. Plant Disease.* **68** (8) : 718-719.

Jiang, Y.M. and Chen, F 1995 A study of polyamine changes and browning of fruit during cold storage of Litchi (*Litchi chinensis* Sonn.) *Postharvest Biology and Technology*, 5 : 245-250.

Joshi, G.D. and Paralkar P. S. 1991 Effect of ripening media and storage behaviour of sapota fruits, National Sem. on Optimization and Utilization of Sapota, GAU Navsari (8 Oct. 1991).

Kadkol, M.V. and Gopalkrishnan, K.S. 1971. Comparative efficacy of Aureofungin in the Control of Downy Mildew of Grapes. *Indian Phytopath.* **24** (3) : 495-499.

Kadambi, K. 1954. On Loranthus control. *Indian Far* **80** : 493.

Kaiser, C. 1995 Litchi (*Litchi chinensis* Sonn.) pericarp colour retention- alternatives to sulphur. *South African Litchi Growers Association Yearbook* **7** : 47-52.

Kamo, T., Hirai, N. Tsuida, M., Fujioka, D., Ohigashi, H. 2000 Changes in the Content and Biosynthesis of Phtoalexins in Banana Fruit. *Bioscience, Biotechnology* and *Biochemistry* **64 (10)** 2089-2098.

Kanapathipillai, V.S. 1988 Non-chemical Control of *Colletotrichum musae*, the anthracnose pathogen of bananas. *Proc. 5th Int. Congr. Plant Pathol,* Kyoto, Japan 4-22.

Kanitkar, U.K. and Uppal, B.N. 1939. Twig Blight and Fruit Rot of Mango. *Curr. Sci.* **8** : 470-471.

Kapoor, I.J. and Tandon, R.N. 1969a. Post-infection changes in organic acids of tomato fruits caused by *Drechslera australiense. Indian Phytopath.* **22** (3) : 408-410.

Kapoor, I.J. and Tandon, R.N. 1969 b. *Naturwisen schaften* **56** : 40-41.

Kapoor, I.J. and Tandon, R.N. 1970. A new species of *Macrophoma* causing fruit rot of Guava (*Psidium guajava*). *Indian Phytopath.* **23** : 122-125.

Kapur S.P. Dhillon, R.S. Cheema S.S. and Deol, I.S. 1973 Exocortis-a Virus Disease of Citrus in Punjab, *Punjab Hort. J.* **13** : 221-222.

Kapur, S.P., Cheema, S.S. and Singh, M.P. 1975. Occurrence and Control of Powdery Mildew of Ber (*Zizyphus mauritiana* Lam.) in Punjab. *Jour.* of *Res. Punjab Agri. Univ.* **12** (1) : 26-29.

Kapur, S.P. and Chohan, J.S. 1968. *Pb. Agric Univ. J. Res.* **5** : 56-57.

Kapur, S.P. and Chohan, J.S. 1974. Evaluation of Fungicides for the Control of Fruit Rot of Papaya by *Macrophomina phaseoli. Indian Phytopath.* **27** : 251-252.

Karmarkar, D.V. and Joshi, B.M. 1940. *Indian Fmg.* **1** : 173-177.

Kaur, C. and Maini, S. B. 2001 Health Foods for New Millennium, *Indian Hort.* **45 (4)** : 29-32.

Kaushik, C.D., Thakur, D.P. and Chand, J.N. 1972. Parasitism and Control of *Pestalotia psidii* causing cankerous disease of ripe Guava fruits. *Indian Phytopath.* **25** : 61-64.

Kaushik, R.A. 2001 Waste Utilization in Bael. *Indian Horticulture* **46** (2) : 17

Kehri, H.K. and Chandra, S. 1986. Control of Botryodiplodia Rot of Guava with homoeopathic drug. *Nat. Acad. Sci. Letters* **9** : 301-302.

Khader, Sesa ; Srivastava, K.C. and Pal, R.N. 1989. Effect of Defoliation of ringing on panicle development and fruit set in Mango. **Agr. Res. Dig.** (Karnal), **9** (2) : 81-83.

Khan, M.D. and Khan, A.H. 1958. Some Studies on Mango Malformation of Mango Inflorescence in West Pakistan, *Pak. Assoc. Adv. Sci.* 10th Conf. Agric. Section : 33.

Khan, M.D. and Khan, A.H. 1962. Relation of Growth to Malformation of Inflorescence in Mangoes. *West Pakistan J. Agric. Res.* **1** : 51.

Khan, I. Misra R. S. and Srivastava, R. P. (1976) Effect of Plant Growth Regulaters on the fruit drop and quality of Litchi cultivar, Rose scented *Progressive Horticulture*, **8** : 61-69.

Khanna, K.K. and Chandra, S. 1976. Studies on Storage Disease on Fruits and Vegetables IV. Control of guava fruit rot caused by *Pestalotia psidii* and *Colletotrichum gloeosoporioides* with Aretan. *Curr. Sci.* **45** : 270.

Khanna, K.K. and Chandra, S. 1977. Control of Guava Fruit Rot caused by *Pestalotia psidii* with homoeopathic drugs. *Plant Dis. Reptr.* **61** : 362-366.

Khanna, K.K. and Chandra, S. 1978. A homoeopathic drug control of mango fruit rot caused by *Pestalotia mangiferae* Henn. *Experientia* **34** : 1167-1168.

Khanna, K.K. and Chandra, S. 1989. Further investigations on the control of storage rot of Mango, Guava and Tomato Fruits with Homoeopathic drugs. *Indian Phytopath.* **42** : 436-440.

Klotz, L.J. 1973 *Color Handbook of Citrus Diseases*, Univ. Calif. Riverside, USA, pp. 122.

Kore, S.S. and Bhide, V.P. 1977. New Hosts for *Curvularia lunata* var. *aeria. India Phytopath.* **30** (3) : 409.

Kore, S.S. and Patil, D.S. 1985. Dry Rot Disease of Mango Seedlings caused by *Fusarium soani. Ind. J. Mycol. Pl. Pathol.*, **15** (3) : 287-288.

Koujalagi, C. B. Kunnal, L.B. and Shankara Murthy H. G. 1996 In Bijapur Economics of Pomegranate Cultivation, *Indian Horticulture* **40(4)** 10-11.

Kovacs, G. 1964 Studies on Antibiotic Substances from higher plants with special reference to their plant pathogenic importance. **K. Vet. Hisk Arsskv**. pp. 47-92.

Krishna, A., Prasad, K. and Ojha, N.C. 1986 Antifungal evaluation of leaf extract for the control of some cucurbitaceous fruit rot diseases. *Indian Phytopath*, **39** : 153.

Krishnaiah, J., Satya Prasad, C.H., Singh, T.G. and Thirupathaiah, V. 1985. Post-harvest Protection of Guava Fruits using Decco food grade fruit coatings. *Indian Bot. Reporter,* **4** : 151-153.

Krishnamurthy S. 1979. Studies on internal breakdown in Alphonso Mango (*Mangifera indica* L.). Mango workers' meeting held at Goa, pp 436-439.

Kuc, J. 1987 *Plant Disease Control*, Wiley & Sons, New York, pp. 255-274.

Ladaniya, M. S. 2000 Response of Nagpur Mandarin to Degreening by ethylene trickle method, mechanized wax coating and storage at ambient condition, *J. of Food Sci. & Tech.* (Mysore) **37 (2)** : 103-110.

Lakshmana, Reddy, T.V. 1999 Effect of Pre-harvest Foliar Application of Ca on the Post-harvest Behaviour of Sapota (cv. Kalipatti) fruits. *South Indian Horticulture.* **47** (1/6) 147- 148.

Lal, B. and Arya, A. 1982. Three new fusarial rots of papaya. *Nat. Acad. Sci. Letters* **3** (3) : 73-74.

Lal, B. and Arya, A. 1982. A soft rot of grapes caused by *Phomopsis viticola. Indian Phytopath.* **35** (2) : 261-264.

Lal, B. Arya, A. Agarwal, R. and Tewari, D.K. 1985. Biochemical Changes in Guava Fruit infected with *Phomopsis psidii. Acta Bot. Indica* **13** : 124-126.

Lal, B., Arya, A. and Rai, R.N. 1981. A new soft rot of 'Ber' caused by *Phomopsis natsume* and its chemical control. *Indian Phytopath.* **34** (3) : 361-363.

Lal, B. Arya, A. and Rai, R.N. 1982. A new soft rot of 'Aonla' caused by *Phomopsis phyllanthi* Punith and its chemical control. *Nat. Acad. Sci. Letters* **5** (6) : 183-185.

Lal, B. and Rai, R.N. 1980. Soft-rot of Chiku (*Achras sapota* Linn.) caused by a new species of *Phomopsis. Nat. Acad. Sci. Letters,* **3** : 139-40.

Lal, B. Rai, R.N. and Arya, A. 1979. Colletotrichum Rot of Papaya. *Nat. Acad Sci. Letters* **3**: 259-260.

Lal, B. Rai, R.N. Arya, A. and Tewari, D.K. 1980. A new soft rot of Guava *Nat Acad. Sci. Letters* **3** : 259.260.

Lal, B. Rai, R.N. Arya, A. and Tewari, D.K. 1980 a. A new rot of *Psidium guajava* L. *Nat. Acad. Sci. Letters* **3** : 361-362.

Lal, B. Rai, R.N. Arya, A. and Tewari, D.K. 1982. Chemical Control of Thielaviopsis rot of Guava. *Indian J. Hort.* **39** : 293-294.

Lal, B. Tewari, D.K. Agarwal R. and Srivastava, R.C. 1984. Some new host records of *Ulocladium chartarum* (Prreuss) Simmons. *Proc. Nat. Acad. Sci* India **54** (B) : 129.

Lal, B. and Srivastava A. 2002 Plants as Biopesticides, In : *Plant Pest Management* (ed.) Trivedi, P.C. Pub. by Avishkar Pub. Distributors, Jaipur, pp. 100-127.

Lal T.B. 1953 Are our citrus orchards free from quick decline. *Pl. Prot. Bull.* **5** : 91-94.

Laxminarayana, P. and Reddy, S.M. 1975. A New Post-harvest Disease of Mango. *Indian Phytopath.*, **28** (4) : 529-530.

Lehoczky, J. and Mako, Sz. 1983. Premature Decay of Grapevine caused by *Stereum hirsutum. Kertgazdazag* **15** (131) : 53-66. (Hungary).

Leu, L.S. and Wu, H.G. 1983. Uredospore Germination Infection Colonization of Grape rust fungus, *Phalcospora ampelopsidis. Plant Protection Bull, Taiwan* **25** (3) : 167-175.

Lim, T.K. and Razak, A.R. 1986. Studies on a Phomopsis rot of bagged guava (*Psidium guajava*) fruits in Malaysia. *Fitopatologia Brasileira* **11** (1) : 227-236.

Lin. K.J. and Li. G.X. 1985. Observations on Symptom Development of Jujube Witch's Broom Disease. *Shanxi fruit trees*, **19** (2) : 35-37.

Lonsdale J.H. 1988 Post-harvest Diseases of Litchis (*Litchi chinensis* Sonn.) and their control. *South African litchi growers' association yearbook,* 1 : 16-18. Luttrell, E.S. 1953. Metanconium Leaf and Stem Fleck of Grapes. *Phytopathology* **43** : 347-348.

Luttrell, E.S. and Murphy, M.M. 1953. Effect of Spraying on incidence of diseases and yields of Muscadine Grapes, *Phytopathology*, **43** : 625-633.

Maheshwari, S. K. and Singh 1999 Effect of Temperature and Relative humidity on development of powdery mildew of Ber (*Zizyphus mauritiana). Annals of Plant Protection Sciences.* **7(1)** : 105-107.

Majumdar, P.K. and Sinha, G.C. 1972. Seasonal Variation in the Incidence of Malformation in *Mangifera. Indica* L. *Acta Hort.* **24** : 221.

Majumdar, P.K., and Sinha, G.C. and Singh, R.N. 1970. Effect of exogenous application of alfa naphthyl acetic acid in mango (*Mangifera indica* L.) malformation. *Indian J. Hort.* **27** : 130.

Malhi C.S. and Kaur T. 2001 Birds Damage in Guava Orchards with special refence to roseringed parakeet (*Psittacula krameri*) *Pestology,* XXV **(8)** 47-54.

Malik, P.C. 1963, Mango Malformation-Symptoms, Causes and Cure. *Punjab Hort. J.* **3** : 292.

Malo, S.E. and McMillan, Jr. R.T. 1972. A disease of *Mangifera indica* L. in Florida similar to Mango Malformation. *Florida State Hort. Soc.* **85** : 264.

Mandels, G.R. 1953. Metabolism of sucrose and related oligosaccharides by spores of the fungus, *Myrothecium verruearia, Plant Physiology* **29** : 18-26.

Marlatt, R.B. Robert, J.K. Jr. and Goldweber, S. 1970. Verticillium Wilt of Mango (*Mangifera indica* L.) in Florida, *Pl. Dis. Reptr.* **54** (7) : 569-571.

Marlatt, J.J., Bledsoe, A.M. Gufler, W.D. Luvisi, D.A. 1986. Control of *Botrytis cinerea* on Grape Berries during postharvest storage with reduced levels of sulfurdioxide. *Plant Disease* **70** (11) : 1050-1052.

McDougall, J.C.1940. Report of the Working of the *Dept. Agric.*, in the Central Provinces and Bihar for the year ending 31st March, 1940, 30 p.

Maruta T. 1977 Studies on the Postharvest Physiology and Storage of Citrus Fruits VII Acid metabolism in satsuma mandarin fruit during storage, *Journal of the Japanese Society for Horticultural Science,* **46** : 283-287.

Mathur, M.M. 1996 Cultivating Ber in IGCP of Rajasthan, *Indian Hort.* **41 (2)** : 22.

Mayudi M.V. 1972. *Pseudomonas viticola* sp. nov. Incitant of a new bacterial disease of grapevine. *Phytopath Z.* **73 :** 183-186.

McElroy, W.D. 1961. *Cellular Physiology and Biochemistry*. Prentice-Hall Inc. Englewood Cliffs, New Jersey, 120 pp.

McGrew, J.R. and Still, G.W. 1972. Control of Grape Diseases and Insects in the Eastern United States. *U.S. Dept., Agric. Farmers Bull.* No. 1893.

McMillan, R.T. Jr. 1971. Fungicides for Disease Control on Avocados, Lime and

Mangoes in Florida. *Proc. Fla. Sta. Hort. Soc,.* **84** : 290-294.

McRae, 1924. Economic Botany Part III. Mycology Annual Report Board Scientific Advice, India. 1922-23 : 31-35.

Meah M. B. 2000 Residues of Mancozeb and Iprodone in Mango Fruits treated for the control of stem-end rot. *Pakistan J Agric. Res.* 16(1) 35-39.

Mehrotra, R.S. 1985. *Plant Pathology,* Tata McGraw Hill Pub. Co. Ltd., New Delhi, 771pp.

Mehta, P.R. 1951. Plant Prot. Bull. New Delhi. **3** : 7-12.

Menten, J.O.M., Hassuike, J.T. and Galli, M.A. 1997 Fungicide Evaluation for the Control of Mofocinzento (*Botrytis cinerea*) on grapes. *Ecossistema*. **22** : 41-43.

Meredith, D.S. 1963. *Trans. Brit. Mycol. Soc.* **46** : 537-40.

Meza N. Manzano J. 2000 Storage of Sapodilla Fruits cv Victoria at different temperatures, **Proc. of Int. Soc. Tropical Horticulture, 42.** 236-245.

Minessy, F.A. Biely, M.P. and El-Falil, A. 1974. Effect of Iron Chelates in correcting malformation of terminal bud growth of Mango in Sudan, *Egyptian J. Hort.*, **1** : 212.

Ming J.Y. Xian, L. S., Fang, C. Bio, L. Y. Donglin, Z. 1997 The Control of Post-harvest browning of litchi fruit by sodium bisulphite and hydrochloric acid. *Tropical Science* **37(3)** : 189-192.

Misra, A.P. and Singh, R.P. 1962. Effect of Temperature and Humidity on the development of banana anthracnose. *Indian Phytopath.* **15** : 11-13.

Mishra, B. Prakash, Om, Misra, A.P. 1974. *Pestalotia menezesiana* on Grape Berries from India. *Indian Phytopath,* **27** : 257-258.

Mitra, A. 1935. *Indian J. Agric. Sci.* **5** : 632-637.

Mohanraj, D., Vidyasekaran, P. and Kandaswamy, T.K. 1972. Possible Role of Sugars in the anthracnose disease resistance mechanism in grapevine varieties. *Indian Phytopath*, **25** (1) : 138-139.

Mitra, S.K. 1988 Other fruits, In : Mineral Nutrition of Fruit Crops, Pub. by Naya Prokash, Kolkata 432-441.

Moheam, N.M.A., El-Sharaya, A.S. Badreldin, M. M. 1988, Oesterogen Control of Pomegranate Seeds. *Journal of Chromatography* **438** : 438-442.

Moller, W.J. 1981a. Veritcillium wilt. In : ''*Grape pest management*'' (Eds. D.L. Flaherty, F.L. Jensen, A.N. Kasimatis, H. Kido and W.J. Moller) Publication No. 4105. Univ. Calif. Berkeley, U.S.A. pp. 82-83.

Moller, W.J. 1981b. Phytophthora Crown and Root Rot. In ''Grape Pest Management'' (Eds. D.L. Flaherty, F.L. Jensen, A.N. Kasimatis, H. Kido and W.J. Moller), Publication No. 4105, Univ. Calif. Berkeley, U.S.A. p. 81.

Moller, W.J. 1981c. Armillaria Root Rot. In : ''Grape Pest Management'' (Eds. D.L. Flaherty, F.L. Jensen, A.N. Kasimatis, H. Kido and W.J. Moller, Publ. No. 4105, Univ., Calif, Berkeley, U.S.A. pp. 77-78.

Moller, W.J. and Kasimatis 1981. Eutypa die back of grapevines In : ''Grape Pest Management'' (Eds. D.L. Flaherty, F.L. Jensen, A.N. Kasimatis, H. Kido and W.J. Moller) Publication No. 4105, Univ. Calif. Berkeley U.S.A., pp. 57-61.

Mordue, J.E.M. 1971. *Glomerella cingulata* C.M.I. Description of Pathogenic Fungi and Bacteria No. 315.

Moreira S. 1942 Observacces sobre a tristeza dos citrus on Podridao das radicelas. *Biologico* **8** : 269-72.

Mukherjee S., Paliwal K., Gupta N.K. 1999, Effect of pre-harvest application of some fungicides on the shelf life of ber (*Zizyphus mauritiana* Lamk.) Fruits c.v. Umran. *Indian Hort.* 47 : 1-6.

Mukherji, S.K. and Bhattacharya, S.K. 1965. A New Disease of *Mangifera indica* L. in India caused by *A. tenuissima Pl. Dis. Reptr.*, **49** (5) : 405-407.

Munjal, R.L. and Kapoor, J.N. 1973. *Carica papaya* a new host of *Sphaerotheca fuliginea. Indian Phytopath.*, **26** (2) : 366-367.

Munjal, R.L. and K.K. Sethi, 1966. *Cereospora* species causing leaf spot of grapes in India. *Indian Phytopath.*, **19**(2) : 209-214.

Mundkur, B.B. and Thirumalachar, M.J. 1947. *Mycol. Pap. Imp. Mycol. Inst.* **16** : 27.

Murty, V.D. 1983 Investigation on citrus mosaic virus in Andhra Pradesh, Ph. D. Thesis Shri Venkateswara Univ., Tirupati.

Naik, K.C. 1949. South Indian Fruits and their Culture. P. Varada-chury and Co., Madras.

Nair, N.G. 1985a. The fungi associated with bunch rot of grapes in the Hunter Valley, *Aust. Jour. Agric. Res.* **36** : 435-442.

Nair, N.G. 1985 b. Bunch rot of grapes caused by *Botrytis cinerea In* : "Chemicals in the Vineyard" (Eds. D.C. Lester, R.M. Cirami and T.H. Lee). Aust. Soc. Vitic. Oenol. Inc. Glen Osmond, S. Aust. 59-66.

Nair, N.G. Emett, R.W. Parker, F.E. 1987. Programming applications of dicarboximides to control bunch rot of grapes caused by *Botrytis cinerea Plant Pathology*, **36** (2) : 175-179.

Nair, N.G. and Nadtotchei, A. 1987. Sclerotia of *Botrytis* as a source of Primary inoculum for bunch rot of grapes in New South Wales, Australia. *Journal* of *Phytopathology*, **119** (1) : 42-51.

Nallathambi P., Thakore, B.B.L. and Gour, H.N. 2002 *Epicoccum nigrum* Link on *Ziziphus mauritiana* (Lamk). A new record from India. *J. Mycol. Pl Pathol.*, **32 (1)** : 131.

Namba, S. Yamashita, S. Boi, Y., Yora K., Terrai, Y. and Yano, R. 1979. Grapevine leaf roll virus, a possible member of Closteroviruses. *Ann. Phytopath. Soc. Japan*, **45** : 497- 502.

Nand, N. Shanker G. and Srivastava, A.K. 1991, Guava Allahabad Surkha in deep pink inside *Indian Hort.* 36(2) : 4-5. *References* 19

Naqvi, S.A.M.H. 1996 Managing Post-harvest Diseases of Nagpur Mandarin, *Indian Hort.*, **41(3)** 46-49.

Narsimhan, M.J. 1954. Malformation of Panicles in Mango incited by a species of Eriophyes. *Curr. Sci.* **23** (9) : 297.

Nariani, T.K. 1956. Leaf Curl of Papaya. *Indian Phytopath.* **9** : 151-157.

Nariani, T.K. and Seth, M.L. 1962. Role of Eriophyd Mites in causing Malformation disease in Mango, *Indian Phytopath.* **15** : 231.

Nariani, T.K., Sahambi, H.S. and Chona B.L. (1965) Occurrence of tristeza virus in citrus in Northern India, *Indian Phytopath.* **18** : 220.

Nariani, T.K. Rayychaudhuri, S.P. and Sharma B.C. 1968 Exocortis in Citrus in India. *Pl. Dis. Reptr.* **52(11)** 834.

Nayar, K.K. Ananthakrishnan, T.N. and David, B.V. 1981. *General and Applied Entomology*, T.M.H. Publ. Co. Ltd., New Delhi.

Nene Y.L. and Thapliyal, P.N. 1965 Antifungal Properties of *Anagalis arvensis* extracts *Naturwisen* **52** : 89-90.

Nijjar G.S. 1969 *J. Res.* Ludhiana 6 : 848-858.

Nishimura, S. Kohmota, K. and Otani, H. 1974. *Rev. Plant Res.* **7** : 21-32.

Nogueira, E. M. De C. Feurari J.T., Figuerido, M.B. Namekata, T. 1997 A new disease of *Carica papaya* in Brazil caused by *Leveillula taurica. Arquivos do Instituto Biologico* (Sao Paulo) **64(2)** 149-152.

Norman, C. 1988 EPA sets new policy on pesticide cancer risks. *Science* **242** : 366-367.

Ocfemia, G.O. and Agati, J.A. 1925. The Cause of Anthracnose of Avocado and Mango in Philippines island. *Philipp. Agric.* **14** : 199-216.

Olaya, C.I. 1985 The Wondrous Papaya. *Americas* (Jan- Feb), 15-19.

Oosthuyse S. A. 2000 Cost reduction of powdery mildew control in mango with mono-potassium phosphate. In Proc. 6th Int Symp. on Mango, Thailand, *Acta Horticuturia* **509** : 719-723.

Oprea M. Rafaila, C. and Ionita, A. 1984. Dieback of Grapevine caused by *Eutypa armeniacae* (Hansf.) Carter. *Buletinul de Protectia Plantelor* **1** : 14-17 (c.f. *Review Plant Path.* 1987, 66 : 2956).

Osipyan, L. L. and Batikyan A. G. 1983 Toxicity of the Mould *Penicillium reticulosum* contaminating pomegranates during storage. *Biologicheski Zhurnal Armenii* **36** : 1083-1085.

Pal, R.N. and Chadha, K.L. 1980. Black-tip Disorder of Mango—a Review. *The Punjab Hort. J.* B **20** : 112-121.

Palo, M.A. 1933. Sclerotium seed rot and seedling stem rot of mango. *The Philip. J. Sci.* **52** (3) : 237-261.

Palti, Pinkas, Y. and Mathilda, C. 1974. Powdery Mildew of Mango. *Plant Dis. Reptr.* **58** : 45-49.

Pandey, R.R., Dwivedi, R.S. 1985. *Fusarium oxysporum* f. sp. *psidii* as a pathogen causing wilt of guava in Varanasi district, India. *Phytopath. Z.* **114 (3)** : 243-248.

Pandey, R.S. Bhargava, S.N. and Shukla, D.N. 1980. Two new fungicides for the control of Aspergillus rot of mango. *Nat. Acad. Sci. Letters* (India), **3** : 263-264.

Palazon, I.J. and Palazon C.F. 1988 Study of Antagonism between *Trichoderma viride* Tul. and different species of fungi causing pome fruit rots (cf. *Review of Plant Pathology* 1990 no. 637), p. 73.

Pandey, R.S. Bhargava, S.N. and Shukla, D.N. 1983. Control of Botryodiplodia Rot of Mango by Oil-soap mulsion. *International J. Pl. Path.* **2** : 1983.

Pandey, R.S. Bhargava, S.N. Shukla, D.N. and Dwivedi, D.K. 1983. Two New Fruit Rot Diseases caused by *Alternaria alternata. Indian J. of Trop. Pl. Diseases* **1** : 23.

Pandey, R.S. Bhargava, S.N. Shukla, D.N. and Khati, D.V.S. 1983. Control of

Pestalotia Fruit Rot of Guava by leaf extracts of medicinal plants. *Revista Mexicana de Fitopathologia*, **2** : 56.

Pandey R.S. Khati, D.V.S. Shukla, D.N. and Bhargava, S.N. 1980. A new Phoma rot of Bel fruit. *Indian Phytopath* **33** (2) : 319.

Pandey, R.M. 1993 More Hybrids of Mango, *Indian Hort.* **38**(3) : 4-5.

Pandey, S.C. and Mohammad, A. 1974-75. Studies on Fruit Rot of Mango IIIBrown spot caused by *Pestalotia mangiferae Butl. Proc. Bihar Acad. Sci* **22-23**: 89-95.

Panwar, K.S. and Yas, N.L. 1974. *Cladosporium oxysporum* causing fruit rot of *Punica granatum, Zizyphus jujuba* and *Capsicum annuum. Indian Phytopath.*, **27** : 121-122.

Pareek, O.P. 1997 Arid Horticulture : Its Potential. *Indian Hort.* **42(2)** : 42-46.

Patel C. B. and Jhala, R.C. 1991 Insect pest problem of sapota with special reference to South Gujarat conditions-A review. A paper presented in National Sem. on Optimization and Utilization of Sapota, GAU, Navsari (8 Oct. 1991).

Patel, G. T. 1981 Biology and Chemical Control of Chickoo Bud Borer, *Anarsia* sp. (Lepidoptera : Gelechidae) M.Sc. (Ag.) Thesis submitted to GAU Sardar Krushinagar.

Patel M.K. and Padhya A. C. 1964 Sodium arsenite, copper sulphate spray for the control of citrus canker *Curr. Sci.* **33** : 87-88.

Patel, M.K. Kamat, M.N. and Bhidey, V.P. 1949, Fungi of Bombay Supplement *I. Indian Phytopath*, **2** : 142-145.

Patel, M.K. Kamat, M.N. and Padhye, Y.A. 1950., *Bacillus subtilis* causing soft rot of mango. *Indian Phytopath.*, **3** : 153-154.

Patel, M.K., Kulkarni, Y.S. and Moniz, L. 1948a. *Pseudomonas mangifera indicae* L. pathogenic on mango. *Indian Phytopath.* **1** : 147-152.

Patel, M.K. Moniz, L. and Kulkarni, Y.S. 1948 b. A new bacterial disease of *Mangifera indica* L. *Curr. Sci.*, **27**(6) : 189-190.

Patel, K.D. and Pathak, V.N. 1993 Incidence of Guava Fruit Rots and Losses due to *Rhizopus* and *Botryodiplodia* in Udaipur and Ahmedabad Markets. *Indian J, Mycol. Pl. Pathol.* **23** : 273-277.

Pathak, V.N. 1980. *Diseases of Fruit Crops.* Oxford & IBH Pub. Co., New Delhi.

Pathak, V.N. Goyal, J.P. and Bhatnagar, L.G. 1976. Effect of chemicals and hot water treatment on *Fusarium* and *Rhizopus* rots of papaya. *Indian Phytopath.* **29**(2) : 210-211.

Pathak, V.N. and Shekhawat, P.S. 1976. Efficacy of some fungicides and hot water in control of anthracnose and Aspergillus rot of mango fruits., *Indian Phytopath.*, **29**(3) : 315-317.

Paull, R. R. 1997 Pineapple, In : Postharvest Physiology and Storage of Tropical and Subtropical Fruits (ed. S. K. Mitra), CAB International Pub. New York, 123-143.

Peshney, N.K. Moghe, P.G. and Gahukar, K.B. 1984. Black Heart of Banana fruits caused by *Fusaruim moniliforme. Indian Phytopath.* **37**(4) 682-683.

Popenoe, W. 1920 *The Litchi : Manual of Tropical and Subtropical Fruits,* McMillan Company, New York, pp 312-25.

Peterson, R.A. 1987: Mango Diseases In : First Australian Mango Research Workshop. *Proc. Melbourne, Australia* C.S.I.R.O. 233-247.

Pine, T.S. 1958. Etiology of the Dead-arm Disease of Grapevines. *Phytopathology* **48** : 192-197.

Pole Evans, I.B. 1934. Aiming at better pastures and field crops. Annual Report of the Division of Plant Industry. *Fmg. S. Afr.* IX 105 : 539-568.

Pradhan, S., Jotqani, M.G. and Rai, B.K. 1952. The neem seed deterrent to lucost, *Indian Farming* **12** : 7- 11.

Prakash, O. 1978. A new mango disease and its cure (Abs.) Symposium on plant disease problems held at Jaipur. Oct. 1-3 : 30-31.

Prakash, O. 1979. Chemical Control of Phoma Blight in Mango.*Indian* J. *Mycol. and Pl. Pathol.* **9**(2) : 184-185.

Prakash, O. Mishra, B. and Mishra, A. P. 1974. *Greeneria fuliginea* Scibner and Viola causing bitter rot of grapes. *Indian Phytopath.* **27** : 605-606.

Prakash, O. and Raoof, M.A. 1985a. Spray Schedule for the control of powdery mildew disease of mango, *Indian J. Pl. Pathol.*, **3 (1)** : 51-52.

Prakash, O. and Raoof, M.A. 1985b. Die back disease of mango and its control. IInd International Symposium on Mango, Bangalore, 20th May, 60 p.

Prakash, O. and Raoof, M.A. 1985c. Blossom Blight Disease of Mango. *Indian J. Plant Pathol.* **3 (2)** : 271-272.

Prakash, O. and Raoof, M.A. 1985d. New Records of Fungi on Leaves and Twigs of Mango (*Mangifera indica* L.) *Indian J. Plant Pathol.* **3(2)** : 242-244.

Prakash, O. and Raoof. M.A. 1985e. Bacterial Canker in Mango (Abs,) IInd International Symposium on Mango, Bangalore, 20-24th May, 59 p.

Prakash, O. and Singh, S.J. 1975. *Indian Phytopath.* **28** : 265-266.

Prakash, O. and Singh, U.N. 1976a. New Diseases of Mango. *Fruit Res. Workshop, Hyderabad,* May 24-28.

Prakash, O. and Singh, U.N. 1976 b. Basal Rot of Mango Seedlings caused by *Sclerotium rolfsi. Indian. J. Mycol. & Pl. Pathol.* **6**(1) : 75.

Pradash, O. and Singh, U.N. 1977. Phoma Blight-a new disease of Mango (*Mangifera indica* L.) *Plant Dis. Reptr.* **61** : 419-421.

Prakash, O. and Singh, U.N. 1979. Fungicidal Control of Red Rust of Mango. *Indian J. Mycol. and Pl. Pathol.* **9** (2) : 175-176.

Prakash, O. and Srivastava, K.C. 1987, *Mango Diseases and Their Management -A World Review.* Today and Tomorrow's Print. and Publ. New Delhi. 175.

Prasad, A. Singh, T.N. 1965. Present Status of Mango Malformation Disease. *Indian J. Hort.* **22** : 254.

Prasad, J.S. and Verma, R.A.B. 1976. Investigation of the Diseases of Papaya IV, Post-infection Changes in the Ascorbic Acid content. *Indian Phytopath.* **29** (1) : 84-85.

Prasad, S.S. and Bigrami, R.S. (1973) Investigations on Diseases of Litchi III Fruit rots and their control by post-harvest treatments, *Indian Phytopath.* **26** : 523-527.

Prasad, S.S. and Verma, R.A.B. 1970. Fungal Diseases of Papaya Fruit in Bihar. *Indian Phytopath.* **23** : 722-724.

Prasad, N., Mehta, P.R. and Lal, S.B. 1952. Fusarium Wilt of *Psidium guajava* L. In Uttar Pradesh (India). *Nature* **169** : 753.

Prasad, R. and Nirvan, R.S. 1965. *Punjab Hort.* **4** : 185-190.

Punithalingam, E. 1975. *Trans, Brit. Mycol. Soc.* **64** : 427-435.

Purkayastha, R.P. 1968. Occurrence of *Fusarium solani* (Mart.) Sacc. on the Stem of *Mangifera indica* L. *Sci* and *Cult.* **34** (5) : 213-214.

Raghunathan, R. Mahadevan, A. and Rangaswami, G. 1966. Biochemical Changes in the Banana Fruit-coat caused by *Gloeosporium musarum* infection, *Indian Phytopath.* **19** (2) : 162-167.

Rai, B.K. and Singh, N. 1967. An Observation on Recovery from Malformation in Mango Sapling. *Curr. Sci.* **36** : 525.

Rai, J.N. 1956.. Stylar end rot of the guava fruit (*Psidium guajava* Linn.) *Proc. Ind.. Acad. Sci.* **43** (B) : 55-61.

Rai, R.N. 1982. Pathological and Physiological studies of certain Fungi causing fruit rot diseases. D. Phil. Thesis, University of Allahabad. 217. pp.

Rai, R.N. and Lal, B. 1984. A new variety of *Chaetosphaeropsis truncata* causing fruit rot of banana. *Indian Phytopath.* **37** (3) : 566-567.

Rai, R.N. Lal, B. and Arya, A. 1982 a. Some New Fruit-rot Diseases. *Indian J. Mycol. Pl. Pathol.* **12** : 264-266.

Rai, R.N. Arya, A. Tewari, D.K. ana Lal, B. 1982b. An Unrecorded Post-harvest Fruit-rot of Guava (*Psidium guajava* L.) *Geobios New Reports* **1** : 134.

Rai, R.N. Arya, A. and Lal, B. 1982 c. *Pestalotiopsis* rot of ber (*Zizyphus mauritiana*) *Indian Phytopath* **35** (4) : 709-710.

Raistrick, H. and Smith, G. 1935. *Biochem. J.* (London) **29** : 606-611.

Raital, R.D. Sohi, H.S. and Manjunath, K.L. 1981. *Pestology* **5** : 11-13.

Ram, H.B. and Date, W.B. 1971. Studies on the Storage Behaviour of two late varieties of mangoes treated with fungicidal wax emulsion at room temperature. *Indian J. Hort.* **28** : 206-209.

Ram, S. Bist, L.D. and Dwivedi, T.S. 1976. Internal Necrosis in Mango. Paper presented at Fruit Research Workshop, held at Hyderabad, 24-28 May.

Ramakrishnan, T. S. 1954 *Common Diseases of Citrus in Madras State*. Pub of Govt. of Madras.

Ramkrishnan, 1981. Studies on Bacterial Canker of Mango (Abs.) Paper presented in Third International Symp. on *Plant Path.*, New Delhi, 15 p.

Ramakrishnan, T.S. and Damodaran, S. 1956. *Proc. Ind. Acad. Sci.* **42** B : 213-222.

Randhawa, G.S. 1967. Grape cultivation has a great future. *Indian Hort.* July-Sept., 42-46.

Rangaswami G. 1957 Development of Resistance to Streptomycin in *Xanthomonas citri* and *X malvacearum. Curr. Sci.* **26** : 185-186.

Rangaswami, G. 1984. *Diseases of Crop Plants in India,* Second Edition, Prentice-Hall of India Pvt. Ltd., New Delhi, 520 pp.

Rangaswami, G. Seshadri, V.S. and Lucy Channamma, K.A. 1970. Fungi of South India. *Univ. of Agirc. Sci.*, Bangalore 193 pp.

Ranjan, S. Govindjee and Laloraya, M.M. 1955. Chromatographic studies on the amino acid metabolism of healthy and diseased leaves of *Croton sparsiflorus Proc. Nat. Inst. Sci. India* **21** : 42-47.

Ranjan, S. and Jha V.R. 1940. The Effect of Ethylene and Sulphur dioxide on the Fruit of *Mangifera indica* L. *Proc. Indian Acad. Sci.* **12** : 267-268.

Rao, D.P.C. and Agarwal, S.C. 1976. Efficacy of Fungicides against Phomposis fruit rot of Guava. *Indian Phytopath* **29** : 345-346.

Rao, N. Sastry, L.V.L. and Parpia, H.A.B. 1967. What nutrition fruits contain. *Indian Hort.*, **11** (4) : 87-95.

Rao, V.G. 1963. Some new records of foliicolous fungi imperfecti from India. *Bull. Bot. Soc. Coll. of Sci.* Nagpur., **42** : 5-57.

Rao, V.G. 1966. An account of market and storage disease of fruits and vegetables in Bombay, Maharashtra. *Mycopath* et *Mycol. Appl.* **28** : 165-176.

Raski, D.J., Goheen, A.C., Lider, L.A. and Meredith, C.P. 1983. Strategies against grapevine fan leaf virus and its nematode vector. *Plant Dis. Reptr.* **67** : 335-339.

Rath, G.C. and Mohanan, M.K. 1986. Fungi inciting fruit rot of Mango. *Indian Phytopath.* **39** (4) : 612-613.

Ratnam, C.V. and Nema, K.G. 1967. Studies on Market Diseases of Fruits and

Vegetables. *Andhra Agric. J.* **14** : 60-66.

Rawal, R.D. 1991 Diseases of Sapota and their Management. A paper presented in National Sem. on Optimization and Utilization of Sapota, GAU Navsari (8 Oct. 1991).

Raychaudhuri, S.P. and Lele, V.C. 1967. Diseases of Fruit Crops other than viruses and their control. *Indian Hort.* **11** (4) : 58-63.

Reddick, D. 1909. Necrosis of Grapevine. *Cornell Univ. Agr. Expt, Sta. Bull.* **63** : 323-342.

Reddy, G.S., Govinda Rao, P. and Paparao, A. 1961. Black-banded Disease of Mango. *The Andhra Agric. J.* **8** (2) : 120-123.

Reddy, G.S. and Govinda Rao P. 1961 Is there Tristeza in Andhra Pradesh. In :Price, W.C. (Ed.) Proc. 2nd Conf., Int. Orgn. Citrus Virol. Univ. Florida Press, Gainesville, pp. 132-135.

Reddy G.S., Reddy M. M., Rao, D.M. and Reddy, P.S.N. 1996 Controlling Diseases on Rainfed Ber in Andhra Pradesh. *Indian Hort.* **41(3)** 39-41.

Reddy, G.S. Moses, G.J. Murty, V. D., Reddy, K. D. and Reddy V.R.K. 1972 Occurrence of mosaic and infectious variegation of sweet orange in Andhra Pradesh. Abstr. 3rd Int. Symp. Subtropical and Tropical Hort., Bangalore. p. 130.

Reddy, M.S. and Reddy, K.R.C. 1983. Greeneria Fruit rot-an Endemic Disease of Grapes in India, *Indian Phytopath.* **36** (1) : 110-114.

Reddy, S.M. 1975. Some new leaf spot diseases caused by Hyphomycetes. *Proc. Nat. Acad. Sci. India* **45** B : 97-100.

Reddy, S.M. and Laxminarayana P. 1978. In Physiology of Parasitism Symp. (eds. Agarwal G.P. and Bilgrami K.S.) Today & Tomorrow Print and Publisher, New Delhi, 113-146.

Rishi, N. and Raychaudhuri, S.P. 1981 Some studies on bushy stunt disease of *Grewia asiatica* Linn. Proc. XVII IUFRO World Congress Kyoto, Japan pp. 347-350.

Roth, G. 1963 Post-harvest Decay of Litchi Fruit. *Citrus and Subtropical Fruit Research Institute Nelspruit, Technical Communication,* **11** : 1-16

Rothwell, A. 1975. A Revised List of Plant Diseases in Rhodesia-Additions 1966-72., *Kirkia* **10** (1) : 295-307.

Roy, R.S. and Sharma, C. 1952. *Ind. Hort.* **9** : 39-52.

Roy, S.K. 2001 Strategic Postharvest Management of Fruits and Vegetables, *Indian Hort.* 45 (4) 4-7.

Rubin, B.A. and Artisikhovskaya, E.V. 1964. *Ann. Rev. Phytopath,* **2** : 157-178.

Ruehle, G.D. 1943. A study of the diseases of Avocado and Mango and development of control measures. *Fla. Agric Exp. Sta. Ann. Rept.* 177-178.

Ruechle, G.D. and Ledin, R.B. 1955. Mango Growing in Florida, *Fla. Agric. Exp. Sta. Bull.* 574.

Saccardo, P.A. 1902. Sylloge Fungorum **16** : 1-1219.

Saccardo, P.A. 1906. Sylloge Fungorum **18** : 572.

Sadhu, M.K. Ghosh, S.K. and Bose, T.K. (1975) *Indian Agric.* **19** : 319-24.

Sall, M.A. Teviotdale, B.L. and Savage, S.D. 1981. Bunch rots. In ''Grape Pest Management'' (Eds. D.L. Flaherty, F.L. Jensen, A.N. Kasimatis, H. Kido and W.J. Moller) Publication No. 4105, Univ. Calif. Berkeley, U.S.A. pp. 51-56.

Sandhu G.S. and Saran C.S. 1980 New Records of Lepidoptera on Sapota. *F.A.O. PL. Prot. Bull.* **28**(1) : 43-44.

Sarkar, A. 1960. Leaf Spot Disease of *Mangifera indica* L. caused by *Pestalotia mangiferae* Butl. *Lloydia* **23** : 1-7.

Sarma T.C. and Bhattacharjee, R.N. 2001 Occurrence of aeromycoflora of banana fruit markets at Guwahati, Nat. Symp. Tropical Mycology in 21st Century, Kolkata. 8-10 Feb., 2001.

Sato, S. Tomiyama, K. Katsui, N. and Masamune, T. 1968. *Ann. Phytopath. Soc. Japan*, **34** : 344-345.

Sattar, A. 1946. *Punjab Fr. J.* **10** : 56-58.

Sattar, A. and Malik, S.A. 1939. Some Studies on Anthracnose of Mango caused by *Glomerella cingulata* (Stonem.) Spauld. Sch. *Colletotrichum gloeosporoides* Penz. *Indian Jour. Agric. Sci.* **1** : 511-521.

Savage, D.S. and Sall, M.A. 1984. Botrytis Bunch Rot of Grapes : Influence of trellis type and canopy microclimate *Phytopathology*, **74** : 65-70.

Scott, K.J., Brown B.I., Chaplin, G.R., Wilcox, M.E. and Bain, J.M. 1982, The Control of Rotting and Browning of Litchi Fruit by hot benomyl and plastic film. *Scientia Horticulture* 16, 253-262.

Schwartz, A. 1968. A New Mango Pest *Farming* in *S. Africa* **9** : 7.

Sechet, M. 1955. Note on the existence of perfect forms in the cycle of *Pestalotia* sp. on fruit trees. *Fruit d' outre Mer*., **10**(7) : 276-277.

Selvaraj, Y, Pal, D.K. Subramanyam, M.D. and Iyer C.P.A. 1982 Changes in the chemical composition of four cultivars of Papaya (*Carica papaya L.*) during growth and development. *Journal Horticulture Science*, **57** : 135-143.

Selvaraj Y., Raja M.E., Rawal K.D. 2000 A Ripening Disorder in Mango Fruits. *Indian J. of Hort.* **57(3)** 183-188.

Sen, P.K. 1943. Further studies on black tip of Mango. *Indian J. Agric. Sci.* **13** : 300-333.

Sen, P.K. Malik, P.C. and Roy, P.K. 1943. Toxic Effects of Gases on Plants. *Sci.* and *Cult.* **9 :** 87-88.

Sethi, K.K. and R.L. Munjal, 1963. Studies on the Nutritional Requirements of *Cercospora viticola* (Ces.) Sacc. (= *Mycosphaerella personata*) Higgins. *Indian Phytopath.* **16** (2) : 185-194.

Shah, N.R. and Arya A. 2000 *Studies on some Fungal Biodeteriogens*, Bhartiya Kala Prakashan, New Delhi, pp. 202.

Shanmuganathan, N. and Emmett, R. 1985-86. Disease Problems associated with grape production in the tropics and sub-tropics. In : *Rev. Trop. Pl. Path.* 2 (Eds. S.P. Raychaudhuri and J.P. Verma) 141-173.

Shanmuganathan, N. and Fletcher, G. 1982. Enzyme-linked immunosorbent assay to detect fan-leaf virus in grapevine grown in containers. *Plant Dis. Reprt.* **66** : 704-706.

Shant, P.S. 1975. Preliminary studies on the effect of Plomofix on mango malformation. *Bangladesh Hort.* **3** : 49.

Sharma, R.C. and Dharam, Vir 1985. New Post-harvest Disease of Grapes, *Indian J. Mycol* and *Pl. Pathol.* **15**(3) : 220.

Sharma, V.P. 1996 Phalsa—a nutritious indigenous fruit, *Indian Hort.* **41(3)** : 9.

Sharma, N.K. Chandel, J.S. Gupta, D. Sharma. R. and Badyala, S.D. 1996 In Himachal Pradesh Problem and Prospects of Litchi Cultivation, *Indian Hort.* **41** : 59-61.

Shatre, R., Mali, B.L. and Bhatnagar, M.K. 1996 Efficacy of plant extracts against xanthomonads., 18th Annual Conference, *Ind. J. Mycol. Pl. Pathol.* 26 (Abstr,) p. 115.

Shea, K.N. 1961. Grapevine Disease *Qd. Agric J.* **87** : 248-253.

Shephard, E.F.S. 1931. Mycological division, *Ann. Rept. Mauritius Dept. Agric.* for the year 1929, p. 8-11.

Shukla, D.N., Bhargava, S.N. and Singh, N.K. 1978. A note on stalk rot of Papaya fruits. *Indian J. Hort.* **35**(3) : 282-283.

Shylesha A.N. Chandra R. Thakur N.S.A. (1998) Guava Stem-borer : A Threat to Growers *Indian Hort.* **43**(2) 10-11.

Siddiqi, M.R. 1974. *Siphinema index* CIH Descriptions of Plant Parasitic Nematodes Set 3, No. 45.

Simmonds, J.H.I 1965. A study of the species of *Colletotrichum* causing ripe fruit rots in Queensland. *Queensland J. Agric.* and *Ani. Sci.* **32** : 437-459.

Sindhu, J.S. and Singh, G. 1981. *Indian J. Mycol. Pl. Pathol* **11** : 105-108.

Singh, A. and Sinha, K.K. 1983. Biochemical changes and aflatoxin production in guava fruit by *Aspergillus flavus* and *A. parasiticus*. *Indian Phytopath.* **36**(2) : 365-366.

Singh, A.P. and Bharagava, S.N. 1976. Presence of some antifungal substances in the skin of guava fruits (*Psidium guajava* L.) *Proc. Nat. Acad. Sci.* India **46**(B) 451-452.

Singh, B. 1954. Studies in the family Loranthaceae, list of new hosts of *Dendropthoe falcata* (L.F.) Ett. its reactions with host, the anatomy on its seedling and mature haustorium. *Agra Univ. J. Res.* **3 :** 301.

Singh, B.N. and Chakravarthi, S.C. 1935. Observation on a disease of mango at Beneras. *Sci.* and *Cult.* **1** : 294.

Singh, B.P. 1971. A new leaf spot disease of Mango (*M. indica*) caused by *Curvularia lunata. Sydowia* **24** (1-6) : 191-192.

Singh, B.P. and Tandon, R.N. 1967. Pathological studies on some isloates of *Alternaria tenusis* causing leaf spot diseases of *Musa paradisiaca*. L., *Psidium guajava* L., *Mangifera indica* L. and *Carica papaya* L. *Proc. Nat. Acad. Sci. India* **37** : 109-116.

Singh, B.P. and Tandon, R.N. 1970. *Indian Phytopath.* **23** : 728-729.

Singh, D. 1967 Cultivation of some less known fruits in India. *Indian Hort.* **11(4)** : 51-54.

Singh, D.B. Sharma, T.V.R.S. Singh V and Bandyopadhyay A.K. 1998 Cultivating Pineapple in Andamans. *Indian Hort.* **43(2)** : 20-21.

Singh, J. and Majumdar V.L. 2002 Fungicidal Management of Alternaria Rot of Pomegranate. *J. Mycol. Pl. Pathol.* 32(1) 134.

Singh, J.P. 1974 The Phalsa, *Indian Hort.* **19** : 15-16.

Singh, G. and Anand, S.P. Chohan, J.S. and Kapoor, S.P. 1976. Evaluation of germplasm collections of grapes against Carcospora leaf spot in the Punjab. *Indian J. Mycol. and Plant Pathol.* **5** : 220-221.

Singh, J.P. and Kainsa, R.L. 1983. Microbial flora of Grapes in relation to storage and spoilage. *Indian Phytopath.* **36**(1) : 72-76.

Singh, J.P. Sharma, Sushil and Yamadagni, R. 1985. Control of post harvest black mould of grapes, *Indian Phytopath.* **38**(3) : 531-532.

Singh, K.K. and Jawanda, J.S. 1961. Malformation in Mangoes, *Punjab Hort. J.* **1** : 18.

Singh, K.K. and Mathur, P.B. 1954. A note on the cold storage of Sapotas (*Achras sapota*) *Ind. Jour. Agri. Sci.* **24** :149-150.

Singh, L.B. 1960. *The Mango.* Leonord Hill (Book) Ltd., London.

Singh, L.B., Singh, S.M. and Nirvan, R.S. 1961. Studies on Mango Malformation : Review, Symptoms, Extent, Intensity and Cause. *Hort. Advance,* **5** : 197.

Singh, N. 1980. Some Physiological and Pathological Studies of *Myrothecium carmichaeli, Alternaria tenussima, Drechslera rostrata* and *Cladosporium oxysporum* causing post-harvest diseases of tomato, field bean, capegooseberry and loquat respectively. *D. Phil thesis, Allahabad University, Allahabad*, India 248 pp.

Singh, K. Kunjadia, P. Reddy, G.V. and Kothari, I.L. 1999 Dynamics of extracellular enzymes during solid state fermentation of banana waste by two *Pleurotus*species (*P. ostreatus* and *P. sajor-caju*). Nat. Sym. On Mycology at 2000 and beyond December 21-23 1999, Vallbh-Vidya Nagar.

Singh, R., Bagde, T.R. and Sharma, V.P. 1981 Nat. Sem. on Trop. and Subtrop. Fruit Crop, Bangalore p. 116.

Singh, R. 1969. *Fruits*, National Book Trust, India, New Delhi, 213 pp.

Singh, R.H. 1968. Cultural and Pathological Studies of Certain Fungi, *D. Phil. thesis, Allahabad Univ., Allahabad*, India. 229. pp.

Singh, R.S. and Prashar, M.P. 1981. *Indian Phytopath.* **34** : 124 (Abstr.)

Singh, S.M. 1956. Control of Mango Malformation Disease. *Sci. Cult.* **21** : 532.

Singh, S.M. 1957. Further studies in the control of Mango malformation disease. *Sci.* and *Cult.* **22**(7) : 394-395.

Singh, S.N. and Singh S.P. 1952 Studies on the pollen storage and longevity of pollens of some fruits and vegetables. Agriculture and Animal Husbandry Department, Uttar Pradesh **2** : 3-11.

Singh, S.P. and Singh, R.K. 1972. Studies on sooty mould of mango (*Mangifera indica* L.) in Bihar (Abs.) Proc. Third International Symposium Subtropical and Tropical Horticulture, Bangalore, 121 p.

Sinha, I.B.P. and Bilgrami. R.S. 1987. Effects of Antibiotics on growth, sporulation, spore germination and germ tube elongation of *Alternaria tenuis* Nees and *Pestalotia mangiferae* P. Henn. In : *Perspectives in Mycological Research,* Vol. I. Today and Tomorrow Printers and Publ. **1987** : 131-138.

Sinha, K.K., Shrivastava, A.K. and Ansari A. A. 1998 Mycotoxin Problems in Fruits and Vegetables. In : Post-harvest Diseases of Horticultural Perishable, (Ed) Sharma, N. and Alam M.M. Pub. by International Book Distributor, pp. 274-284.

Sinha, K.K. and Singh, A. 1982 Aflatoxin producing potentials of *Aspergillus flavus* and natural occurrence of aflatoxins in some fruits and vegetables. *Nat. Acad. Science letters,* **5** : 213-214.

Sivanesan, A. and Holliday, P. 1981. *Guignardia bidwellii* C.M.I. Description of Pathogenic Fungi and Bacteria, No. 710.

Small, W. 1924. Annual Report of the Govt. Mycologist. *Rep. Dept. Agric. Uganda* for the year ending 31st Dec. 1922, 29 p.

Small, W. 1926. On the occurrence of a species of *Colletotrichum, Trans. Birt. Mycol. Soc.* **11** : 112-137.

Smilanick J.L. and Sorenson D. 2001 Control of Post-harvest Decay of Citrus fruits with calcium polysulfide. *Post-harvest Biology and Biotechnology*, **21**(2) 157-168.

Somasekhara, Y.M. and Wali, S.Y. (2000) Survey of Incidence of Pomegranate (*Punica granatum* Linn.) wilt (*Ceratocystis fimbriata* Ell & Halst.) *Orissa J of Hort.*, **28**(2) 84-89.

Smoot, J.J. and Segall, R.H. 1963. Hot water as Post-harvest control of mango anthracnose. *Plant Dis. Reptr.* **47** : 739-742.

Spotts, R.A. and Cervantes, L.A. 1986 Populations, Pathogenicity and Benomyl resistance of *Botrytis* spp. *Penicillium* sp. and *Mucor piriformis* in packing houses. *Plant Dis.* 70 : 106-108.

Sohi, H.S. 1983. Disease of Tropical and Sub-tropical Fruits and Their Control—In : Recent Advances in Plant Pathology (eds. Akhtar Husain, B.P. Singh, V.P. Agnihotri and Kisan Singh), Print House India, Lucknow, 521 pp.

Sohi, H.S. and Prakash Om, 1972. *Indian J. Mycol. and Pl. Path.* **2** : 139-142.

Sohi, H.S. and Prakash, Om . 1978. A New Species of *Pestalotiopsis* on Banana. *Indian Phytopath.* **31** : 252-254.

Sohi, H.S. and Sharma, S.L. Nayar, S.K. Shyam, K. 1972. New Records of Fungi from Himachal Pradesh, V. *Indian Phytopath.* **25**(3) : 439-441.

Sohi, H.S. Sokhi, S.S. and Tiwari, R.P. 1973. Studies on the Storage Rot of Mango caused by *Colletorichum goleosporioides* Penz. and its control. *Phytopath. Mediterranea XII* : 114-116.

Sohi, H.S. and Sridhar, T.S. 1971. *Indian J. Hort.* **27** : 201-204.

Sohi, H.S. and Sridhar, T.S. 1972. *Indian J. Hort.* **28** : 116.

Solangi, G.P. and Malik, M.M.S. 1971 Studies on Physiological behaviour of *Colletotrichum dematium* (Pers. ex Fr.) Grove, causing anthracnose of papaya fruit. *J. of Agr. Res.* Lahore **9** (1) : 50-61.

Spalding, D.H. and Reeder, W.F. 1972. Post-harvest Disorders of Mangoes as affected by Fungicides and Heat Treatments. *Plant Dis. Reptr.* **56**(9) : 751-753.

Spotts, R.A. 1979. Use of Bay Meb 6447 for Eradication of Grape Black Rot caused by *Guignardia bidwelli. Plant. Dis. Reptr.* **63** : 967-969.

Sridhar, T.S. 1974. Evaluation of Fungicides against *Rhizopus* spp. causing soft rot of fruits. *Hindustan Antibiotic Bull.* **17** : 31-34.

Sridhar, T.S. and Sohi, H.S. 1973. Powdery Mildew of Mango and its Control. *Indian Phytopath.* **26**(2) : 361-362.

Srivastava, A.K. and Lal, B. 1997 Studies on Biofungicidal Properties of leaf extracts of some plants. *Indian Phytopath.* 50 : 408-411.

Srivastava, A.K., Singh H.K. and Singh I.S. 1998 Genetic Diversity :Bael (*Aegle marmelos* Correa) *Indian Hort.* 42(4) : cover page ii

Srivastava, H.C. and Mehta P.R. 1951 A new species of *Cercospora* on *Grewia asiatica* Linn. *Indian Phytopath.* 4 : 67-70.

Srivastava, M.P. 1966. Cultural and Pathological Studies of Certain Fungi causing Post-harvest Diseases of some fruits. *D. Phil. thesis,* Allahabad Univ., Allahabad, India 299 pp.

Srivastava, M.P. and Tandon, R.N. Bilgrami, K.S. and Ghosh, A.K. 1964 a. Studies on Fungal Diseases of some tropical fruits and fruit trees. *Phytopath. Z.* **50**(3) : 250-261.

Srivastava, M.P. Chandra, S. and Tondon, R.N. 1964 b. Post-harvest Diseases of Some Fruits and Vegetables. *Proc. Nat. Acad. Sci.* India, **34**(4) : 339-342.

Srivastava, M.P. and Tandon, R.N. 1966. *Experientia* **22** : 789.

Srivastava, M.P. and Tandon, R.N. 1968. Influence of Temperature on Botryodiplodia rot of Citrus and Sapodilla. *Indian Phytopath.* **21** : 195-197.

Srivastava, M.P. and Tandon, R.N. 1969. Symptoms of Post-harvest Diseases of Guava. *Pl. Dis. Reptr.* **53** : 206-208.

Srivastava, M.P. and Tandon, R.N. 1971. Post-harvest Diseases of Banana in India. *Indian Phytopath.* **24**(1) : 115-118.

Srivastava, M.P. and Tandon, R.N. Bhargava, S.N. and Ghosh, A.K. 1965. Studies on Fungal Diseases of Some Tropical Fruits III. Some Post-harvest Diseases of Mango. *Proc. Nat. Acad. Sci.* India **35 B** : 69-75.

Srivastava, R.C., Arora, S., Tewari, D.K. and Lal, B. 1986. Phoma rot of *Phyllanthus emblica* L. *Geobios new Reports*, **5** : 195-196.

Srivastava, R.C. Tewari, D.K., Katiyar, N. and Lal, B. 1988. Efficacy of some fungicides against *Chaetosphaeropsis truncata* var. *indica* Rai & Lal causing banana rot. *Geobios new Reports*, **7** : 94-96.

Srivastava, R.P. 1963a. Effect of smoke and industrial gases on the horticultural plants with special reference to black tip disease of mango. *Gardening,* 24-36.

Srivastava, R.P. 1963 b. The black tip disease of Mango, causes and control. *Punjab Hort.* J. **8** : 226-228.

Stephens S.E. 1955 The Litchi. *Queensland Agriculture Journal* **44** : 191-194.

Stinson, E.E., Osman S.F., Heisler, E.G., Siciliano, J. and Bills, D.D. 1981 Mycotoxin production in whole tomatoes, apples, oranges and lemons., *J. Agric. Food Che.* **29** : 790- 792.

Stevens, H.E. 1936. Control of Mango Blight and Anthracnose. *Proc. Fla. Hort. Soc.* **49** : 125-130.

Stevens, F.L. and Pierce, A.S. 1933. Fungi from Bombay. *Indian J. Agric. Sci.* **3** : 912-916.

Stevens, R.B. and Stevens, N.E. 1952. Diseases in Plant, *Chronica Botanica* Watham, Massachusetts, U.S.A.

Steyaert, R.T. 1953. Pestalotiopsis from the Gold Coast and Togaland. *Trans. Brit. Mycol. Soc.* **36** : 235-242.

Strobel, G.A. and Hewitt, W.B. 1964. Times of Infection and Latency of *Diplodia viticola* in *Vitis vinifera* var. Thompson Seedless. *Phytopathology* **54** : 634-639.

Subramanian, C.V. 1956. Hyphomycetes II. *Jour. Indian Bot. Soc.* **35**(1) : 446-494.

Subramanyam, H. Krishnamurthy, S. Subhadra, N.V. and Dalal, V.B. 1971. Studies on Internal Break-down. A physiological ripening disorder in Alphonso Mangoes (*Mangifera indica* L.) *Tropical Science* **13** : 203-210.

Subba Rao, M.K. 1936. Pink disease. *Bull. Tea Sci. Dept. Unit. Plant Assoc. S. India* **10** : 25.

Suhag, L.S. and Kaushik, J.C. 1982. Studies on four pathogenic fungi on grapevine in North India. *Indian Phytopath.* **35**(2) : 344-346.

Suhag, L.S. and Mehta, N. 1983. Participation of host in modifying the toxicity of some antipowdery mildew chemicals. *Indian Phytopath.* **26**(2) : 402-403.

Suit, R.E. and Ducharme, E.P. 1946. The Cephaleoros Disease of Citrus. *Citrus Ind.* **27** : 3.

Summanwar, A.S. Raychaudhurim S.P. and Pathak S.C. 1966. Association of the fungus *Fusarium moniliforme* Sheld. with the malformation in mango (*Mangifera indica* L.) *Indian Phytopath.* **19** : 227-228.

Su, M.T. 1934. Report of the Mycologist. Burma Mandalay for the year ending 31st March, 1934. *Rept. Dept. Agric.* Burma, 1933-4 : 25-33.

Swarts, D.H. 1989 Die effek van SO_2 behandling op naoesbederfontwikkeling by lietsjies. South African Litchi Growers' Association Year-book **2** : 22-24.

Swinburne, T.R. 1973 The resistance of immature Bramley's seedling apples to rotting by *Nectria galligenia* Bre. In : *Fungal pathogenicity and the Plant' Response.* Academic Press, London, pp. 365-382.

Tanda, S. and Braun, U. 1985. Two new powdery mildew fungi on *Carica papaya and Phyllanthus flexuosus. Trans.* of *Mycol. Soc. of Japan,* **26**(3) : 315-319.

Tandon, D.K. and Kalra S.K. 2001 Ripening early harvested mangoes with Ethrel, *Indian Hort.* **46(2)** : 32-33.

Tandon, I.N. 1959. *Hort. Advance* **3** : 115-112.

Tandon, I.N. and Singh, B.B. 1968. Control of Mango Anthracnose by hot water treatment. *Indian Phytopath.* **21** : 321-326.

Tandon, I.N. and Singh, B.B. 1969. Studies on Anthracnose of Guava and its Control, **22**(3) : 322-326.

Tandon, I.N. Bhargava, V. and Jamaluddin, 1977. Chemical control of decay of fruits of *Vitis vinifera* caused by *A. niger* and *P. citrinum. Indian Phytopath.* **30** (3) : 403-404.

Tandon, M.P. Jamaluddin and Bhargava, V. 1975 a. Some new fruit rot diseases. *Indian Phytopath.* **28**(4) : 570-572.

Tandon, M.P. Jamaluddin and Bhargava, V. 1975 b. *Curr. Sci.* **44** : 478.

Tandon, M.P. and Tandon R.N. (1975) Rot of Fruits of Litchi (*Litchi chinensis*) in marketing processes *Indian Phytopath.* **28** : 530-531.

Tandon, R.N. 1967. Final Technical Report PL-480 Scheme (FG-IN-133), University of Allahabad.

Tandon, R.N. 1970. Certain problems of post-harvest diseases of fruits and vegetables. *Indian Phytopath.* **23**(1) : 1-15.

Tandon, R.N. and Bhargava, S.N. 1962. *Curr. Sci.* **31** : 344-345.

Tandon, R.N. and Bilgrami, K.S. 1961. Some new leaf spot diseases. *Curr. Sci.* **30** : 348.

Tandon, R.N., Sisodia, U.S. and Bilgrami, K.S. 1955. Pathological Studies of *Pestalotia mangiferae. Proc. Indian Acad. Sci.* **42** : 219-255.

Tandon, R.N. Srivastava, M.P. Bhargava, S.N. and Ghosh, A.K. 1965. Studies on fungal diseases of some tropical fruits (Abs.) 34th Ann. Sess. of *Nat. Acad. Sci. India.* 58-59 p.

Tandon, R.N. and Varma, A. 1964. Some new storage diseases of fruits and vegetables. *Curr. Sci.* **33** : 625-627.

Tanne, E. Sela, I. and Harpaz, I. 1974. Transmission of grapevine leaf roll virus to herbaceous plants. *Phytopath. Z.* **80 :** 176-180.

Tanne, E. Sela, I. Klein, M. and Harpaz, I. 1977. Purification and characterization of a virus associated with grapevine leaf roll disease. *Phytopathology* **67** : 442-447.

Tanka, S. 1933. *Agr. Kyoto Imp. Univ.* **28** : 1-31.

Tarmizi, S.A. Malik, T.M.T.A. Pauziah, M. 1988. Harvesting time and their effect on the incidence of fruit rot in mango cv. Harumanis *M A R D I RES. J* **16**(1) : 195-200.

Taya, R.S., Taya J.S. & Singh, J.P. 1999 Studies on the transmission of papaya ring spot virus by *Aphis gossypii* Glover. *Agricultural Science Digest* (Karnal) **19**(1) : 39-42.

Telizm D. Goheen, A.C. and Valle, P. 1980. Occurrence and Spread of grape corky bark and stem pitting in Mexico. *Plant Dis. Reptr.* **64** : 584-586.

Tewari, D.K. Srivastava, R.C. Katiyar, N., Arora, S. and Lal, B. 1988 a. Post-harvest

Diseases of Fruits, *Proc. Nat. Acad. Sci. India* **58**(B) : 345-346.

Tewari, D.K. Srivastava, R.C. Katiyar, N. and Lal, B. 1988 b. Chemical Control of Thielaviopsis rot of Papaya. *Indian Phytopath.* **41** : 491-492.

Tewari, V.P. and Srivastava, J.S. 1977. Occurrence of *Lophodermium mangiferae* on mango leaf in Varanasi. *Curr. Sci.* **46** : 55.

Thakur, D.P. 1972. Factors influencing storage rot of certain fruits and vegetables caused by three species of *Rhizopus. Indian Phytopath.* **25**(3) : 354-358.

Thakur, D.P. and Chenulu, V.V. 1966. Survey for storage and market deseases of certain fruits and vegetables in Delhi (Abs.) *Intern. Symp.* on *Plant Pathology* Dec. 27 New Delhi, 35-36 p.

Thakur, D.P. and Chenulu, V.V. 1970. Chemical control of soft rot of apple and mango fruits caused by *Rhizopus arrhizus. Indian Phytopath.* **23**(1) : 58-61.

Thakur, D.P. and Misra, A.P. 1964. *Indian J. Hort.* **21** : 349-351.

Thakur, D.P. and Misra, A.P. 1966. *Indian J. Hort.* **23** : 59-61.

Thammaiah, N., Hanmashethi, S.I., Shirol, A.M. Kanamadi, V.C. Hulamani, N.C. 1999 Evaulation of Sigatoka leaf spot incidence in high density population of Rajapuri Banana. *South Indian Hort.* **47**(1/6) 168-169.

Thangaraj, T. Irulappan, I. 1988. Effectiveness of hot water treatment, waxing and cool chamber storage in prolonging self- life of mango. *South Indian Horticulture* **36**(6) : 327-328.

Thomas, P. Saxena, S.C., Chandra, R., Rao, R. and Bhatia, C.R. 1993 X-ray imaging for detecting spongy tissue, an internal disorder in fruits of Alphonso Mango (*Mangifera indica) Journal of Horticultural Science* **68** : 803-806.

Tripathi, R.D. 1955. Malformation Disease of Mineral Nutrients. *Indian J. Hort.* **12** : 173.

Transomo, A. 1989 *Trichoderma harzianum* used for biological control of storage rot on carrots. *Norwegian J. Agri.* Sci **3** : 157-161.

Tzeng, H.L. and Goheen, A.C. 1984. Electron microscopic studies on the corky bark and leaf roll virus diseases of grapevines. *Phytopathology* **74** : 1142 (Abstr.)

Underhill, S.J.R. Coats I. M. and Saks Y 1997 Litchi, In: *Post-harvest physiology and storage of tropical and subtropical fruits* (ed.), Mitra S.K. CAB International Pub, London, pp. 191-208.

Ullasa, B.A. and Rawal, R.D. 1975. Occurrence of a new post-harvest disease of mango due to *Pestalotia gradicola* (Abs.) *Second Int. Symp.* on *Mango*, Bangalore, (India) 64 p.

Ullasa, B.A. Sohi, H.S. and Ganapathy, K.M. 1974. Ascochyta leaf spot of papaya and its perfect state. *Ind. J. Mycol* and *Pl. Pathol.* **4** : 218-219.

Upadhyaya K. and Tripathi K.C. 1996 Occurrence of Stylar End-rot of Guava in Garhwal Himalayas. *Progressive Horticulture* **28** (3-4) 150-151. Uppal, B.N. 1937. Summary of Work of Plant Pathologist Bombay, *Rep. Dep. Agric Bombay*. 1935-36 : 203-207.

Uppal, B.N. Patel, M.K. and Kamat, M.N. 1935. The Fungi of Bombay. *Dept. of Agric. Bombay Bull.* **176** : 56 pp.

Uppal, B.N. Patel, M.K. and Kamat, M.N. 1941. Powdery Mildew of Mango. *J. Univ. Bombay*, **9** : 12-16.

United States Department of Agriculture USDA, 1960. Index of Plant Diseases in the United States Agric. Handbook No. 165, U.S. Dept. Agric. Washington, D.C. U.S.A. pp. 482-484.

Vaheeduddin, S. 1953. The Mango. *Dept. of Agric. Hyderabad,* 121 + 11, 23 pp.

Vala, D.G. Solanki, K.U. Desai, V.D. and Joshi, H.U. 1985. Diseases of Mango occurring in Gujarat State (Abs.) *Second International Symp. on Mango,* Bangalore, India, 64-65.

Vanderweyen, A. 1964. Anthracnose of Mango in Morocco. *Rev. Res. Agron Moroc.* **11** : 51-59.

Varma, A. 1983. Mango Malformation Exotic Plant Quarantine Pests and Procedures for Introduction of Plant Materials, 173-188.

Varma, A. Lele, V.C. Majumdar, P.K. Asha Ram, Sahidanand, J. Shukla, U.S. Sinha, G.C. Yadav, T.D. and Raychaudhuri, S.P. 1969. Mango Malformation. Proc. ICAR Workshop Fruit Research, Ludhiana, 28th-30th April, 1969 : 1.

Varma, A. Lele, V.C. and Goswami, B.K. 1974a. Mango Malformation, Current Trends in Plant Path. S.P. Raychaudhury and J.P. Verma (eds.) Bot. Dept. Lucknow University, Lucknow, India, 196-208.

Varma, A. Lele, V.C. Raychaudhuri, S.P. Asha Ram and Asha, Singh, 1974b. Mango Malformation Fungal Disease, *Phytopath Z.* **79** : 254.

Varma, A. Raychaudhuri, S.P. Lele, V.C. and Asha Ram, 1971. Preliminary investigations on epidemiology and control of mango malformation. *Proc. Indian Nat. Acad.* **37 B** : 291.

Varma, A. Raychaudhuri, S.P. Lele, V.C. and Asha Ram, 1972. Towards the understanding of the Problem of Mango Malformation. *Acta Hort.* **24** : 237.

Vasudeva, R.S. and Capoor S.P. 1958 Citrus Decline in Bombay State. *Pl. Prot. Bull. F.A.O.* **6** : 91.

Vedram and Dharam Vir 1986. Efficacy of fungicides and relative evaluation of various post-harvest chemical treatments against spoilage of banana fruit caused by *Curvularia lunata. Indian Phytopath.* **39**(4) : 594-595.

Veera, S. and Das R.C. 1973 Effect of growth regulators on fruit quality in litchi variety Muzaffarpur. *Indian J. of Hort.* 30 : 363-365.

Verma, R.A.B., Prasad, J.S. and Verma, S.K. 1980 Aflatoxin production in fruits under market condition. *J. Indian Bot. Soc.* 59 : 28.

Verma, G.S. 1950. Tip Pulp of Mango Fruit. *Curr. Sci.* **19** : 246.

Verma, G.S. 1952. The formation of lesions by gases on mango fruits. *J. Indian Bot. Soc.* **31** : 316-341.

Vidhyasekaran, P. and Durairaj, P. 1973. Role of auxin phenol complex in shot hole syndrome development in mango incited by *C. gleosporioides. Indian Phytopath.* **26** : 49-55.

Vicent, A., Armengol, J. Sales, R. Garcia Jimenez, J 2000 First report of Alternaria brown spot of citrus in Spain. *Plant Disease* **84**(9) 1044.

Von Loessecke, H.W. 1949 *Bananas.* Inter Science Publ, New York.

Vidhyasekaran, P. and Parambaramani, C. 1971a. Nitrogen metabolism of alga infected plants. *Indian Phytopath* **24**(3) : 500-504.

Vidhyasekaran, P. and Parambaramani, C. 1971b. Carbon metabolism of alga infected plants. *Indian Phytopath* **24**(2) : 369-374.

Vidhyasekaran, P. and Parambaramani, C. 1972. Mineral metabolism of alga infected plants. *Indian Phytopath* **25**(1) : 86-90.

Viegas, A.P. 1946. Some Fungi of Brazil (S. Paulo) *Bragantia* **6** : 1-37.

Viljoen, N.M. and Kotze, J.M. 1972. Bacterial Black-spot of Mango. *Phytophylactica* **4**(3) : 93-94.

Vuittenez, A. 1970. Fan leaf of grapevine In : "Virus Diseases of Small Fruits and Grapevines" (Ed. N.W. Frazer). Univ. Calif. Berkeley, U.S.A., pp. 217-232.

Wadia, KDR and Manoharachary, C. 1982. New Post-harvest Diseases of Jambolan fruits. *Indian Phytopath.* **35**(4) : 707-708.

Wager, V.A. 1937. Mango Diseases in South Africa. *Fmg. S. Africa*, 12-14 NP.

Wangikar, P.D. and Raut, J.G. 1972. Notes on some plant diseases round about Akola. *Punjab Rao Krishi Vidyapeeth Res. J.* **1**(1) : 139-140.

Wahid, O.A.A. 1999 Comparative study on five isolates of *Colletotrichum gloeosporioides* causing guava anthracnose in Egypt and its control. *Microbiological Research* 154 : 63- 69.

Wallace J.M. 1959 A half century of research on Psorosis, In ; Wallace, J.M. (Ed.) *Citrus virus diseases*, Univ. Calif. Berkeley, pp. 5-21.

Wallace, J.M. 1968 Recent developments in the citrus psorosis diseases In : Childe, J.F.L. (Ed.) *Proc. 4th Conf. Int. Orgn. Citrus Virol.* Univ. Florida Press, Gainesville pp. 1-9.

Wallace, J.M. and Draker, R.J. 1972 Studies on recovery of citrus plants from seedling yellows and the resulting protection against infection, In : Price, W.C. (Ed.) *Proc. 5th Conf. int. Orgn. Citrus Virol.* Univ. Florida Press Gainesville pp. 127-136.

Waks, J. Chaltz, E. Schiffmann-Nadel, M. and Lomeniece, E. 1985 Relationship among ventilation of citrus storage room, internal fruit atmosphere, and fruit quality. *Journal of the American Society for Horticultural Science,* **110** : 398-402.

Waraiteh, K.S. 1977. Additions to Pyrenomycetes Fungi of India. *Proc. Nat. Acad. Sci. India. 46th Ann. Sess.*, p. 23.

Wardlaw, C.W. 1972, *Banana diseases including plantains* Abacca 2nd ed. CMI Kew, Surrey Logon Longmans, pp. 878.

Wardlaw, C.W. and Leonard, E.R. 1936. The Storage of West Indies Mangoes. *Mem. Low. Temp. Res. Sta.* **2** : 47.

Watt, G. 1891. Dictionary of the Economic Products of India **5** : 149.

White, R.P. 1931. Diseases of **Delphinium** or Larkspur. *New Jersey Agric. Exp. Sta. Cir.* **2345** : 2-4.

Wilson, C.L. 1989 Managing the Microflora of Harvested Fruits and Vegetables enhance resistance. *Phytopathology* **79** : 1387-1390.

Willett, M.J. Kupferman, E.A., Goberts R.G., Spotts R.A. Sugar, D. Apel, D. and Ewart, H.W. 1992 Integrated management of post-harvest diseases and disorders of apples, pears and cherries. *Tree fruit post-harvest J.* **3** : 3-16.

Wilson, C.L. and Pusey P.L. 1985 Potential for biological control of post-harvest plant diseases. *Plant Dis.* **69** : 375-378.

Wilkie, J.P. Dye, D.W. and Watson, D.R.W. 1973. Further Hosts of *Pseudomonas viridiflava, N. Z. J. S. Agric. Res.* **16** : 315-323.

Williamson, D. and Tandon, R.N. 1965 a. *Naturwissen Schaften* **52** : 166.

Williamson, D. and Tandon, R.N. 1965 b. Some Pathological Studies on *Botryodiplodea theobromae* Pat. causing banana rot. *Mycopath* et *Mycol. Appl.* **29** : 245-253.

Wilson, D.M. and Nuovo, G.L. 1973. *Appl. Microbiol.* **26** : 124-125.

Winkler, A.J. Cook, J.A. Kleiwer, W.M. and Lider, L.A. 1974. General Viticulture. Univ. Calif. Press, Berkeley, U.S.A.

Wisniewski, M., Wilson C.L. and Chalutz E. 1988 Variability in Biocontrol of fruit rots among isolates of the yeast *Debaryomyces hansenii. Phytopathology* **78** : 1592 (Abstr).

Wisniewski, M., Biles, C.L., Droby S., McLaughlin, R. Wilson C.L. and Chalutz E. 1991 Mode of action of post-harvest biocontrol yeast *Pichia guilliermondii* I Characterization of attachment of *Botrytis cineria. Physiological and Molecular Plant Pathology* **39** : 245-258.

Woodhouse, E.J. 1909. The Mango of Bihar. *Quarterly Jan. Dept.* of *Agric* Bengal (India) **2** : 283.

Wong, L.S., Jakobi, K.K. and Giles, J.E. 1991 The influence of hot benomyl on the appearance of stored lychee (*Litchi chinensis* Sonn.) *Scientia Horticultrae* **46** : 245-251.

Xin Ge H., Chang Z. J., Jing Wu 2000 The reasons for the rampant occurrence of citrus anthracnose (*Glomerella cingulata*) disease and its control. *South China Fruits.* **29**(3) 13.

Yadava, A.S. and Udainarain, 1970. Effect of micro succession on the fallen stem of *Mangifera indica* L. II The ability of micro-fungi to cause actual decay of the woody stem *in vitro* and the factors which influence it. *J Indian Bot. Soc.* **47** (3-4) : 274-281.

Yadva I.S. 1997 Mango Research in India : The Past 50 years, *Indian Hort.* **42**(2) 10-17.

Young, T.W. 1957. Soft nose-a physiological disorder in mango fruit. *Fla Stat Hort. Sci.* **70** : 280-283.

Young, T.W. 1958. Mango Research on Sandy Soil. *Proc. 17th Ann. Meet. Fla Mango Forum* 1957 : 26-30.

Yule, Sir Henry, Burnell, A. C. and Hobsen-Jobson 1903 A glossary of Anglo-Indian colloquial words and phrases, etymological, historical, geographical and discursive. New edition, William J. Murray, London.

Zem Zami, M., Gaunsey, S.M., Nadori, E.B. and Hill, J.H. 1999 Biological and Serological Characterization of Citrus Tristeza Virus (TV) isolates from Morocco. *Phytopathologia Mediteranea* **38**(2) : 95-100.

Zauberman, G. and Barkai-Golan R. 1975. *Phytopathology,* **65** : 216.

Index

A

I

J

K

L

M

N

O

P

R

S